Speaking, Stammering, Singing, Shouting

SOUND IN HISTORY

Emma Dillon, Series Editor

A complete list of books in the series is available from the publisher.

Speaking, Stammering, Singing, Shouting

A Social History of the Modern Voice

Josephine Hoegaerts

PENN

UNIVERSITY OF PENNSYLVANIA PRESS

PHILADELPHIA

Published by
University of Pennsylvania Press
Philadelphia, Pennsylvania 19104-4112
www.pennpress.org

Printed in the United States of America on acid-free paper

10 9 8 7 6 5 4 3 2 1

A Cataloging-in-Publication record is available from the Library of Congress

Hardback ISBN 978-1-5128-2773-6
Ebook ISBN 978-1-5128-2774-3

CONTENTS

Speaking, Stammering, Singing, Shouting

Introduction

A History of the "Modern" Voice?

> No sound goes more directly to the soul than the human voice.
>
> —Marc Colombat de l'Isère

In his 1857 *Treatise upon the Disease and Hygiene of the Organs of the Voice*, French speech pathologist Marc Colombat de l'Isère included a short eulogy on the human voice in his otherwise rather dry and technical research report. The human voice, de L'Isère insisted, surpassed any man-made instrument, and was "beyond imitation." His assertion on the effects of the human voice on the soul must have rung true for many of his readers. Connections between the soul and the voice have a long history. In a sense, the *ortophoniste*'s statement was a simple reversal of the older early modern discourse in which the voice, carried by breath, was an outer expression of the soul. In the sixteenth century, the faithful would be warned in church that words—traveling through air toward their ears—could have a profound influence on the state of their soul.[1] If virtuous and benign, such as the words emanating from the mouth of a priest, the voice entering a Christian's ear could open his heart to God. But the unprotected nature of the ears, which unlike eyes cannot be closed off at will, also made them vulnerable to unwelcome intrusions. The devil could easily lisp his way into the unsuspecting listener's ear and gain a hold over his soul as well. In Colombat's discourse, published well after the discovery of sound waves, voice is no longer carried by breath. And, more importantly, the voices gaining access to the soul are no longer seen as otherworldly in nature. They are, explicitly and unmistakably, human.

Although the vocabulary of the soul was slowly disappearing when de l'Isère wrote his treatise, new incarnations of this idea reemerged: the voice was and is heard as a manifestation of temperament, personality, and feeling. Minister and educator Theodore Schmauk, writing at the end of the nineteenth century, listed a number of possibilities:

> In speaking of Quality *with reference to the various properties of tone*, we say a voice is rich, full, deep, piercing, sweet, rough, smooth, ringing, bird-like, flute-like, trumpet-like, manly, womanly, child-like. *As a reflection of the apparent natural mental state of the speaker*, a voice is pathetic, solemn, tranquil, grave, serious, animated, gay, playful, mirthful, rollicking, melancholy, sublime, courageous, scornful, defiant, threatening, despairing, awe-stricken, alarmed, horrified, revengeful, kind, tender, hopeful, truthful.[2]

The connection between soul and voice was cemented in the nineteenth century by the recurring idea that voice and humanity were interconnected.[3] De l'Isère apparently felt the need to specify that only the human voice goes quite so directly to the soul, suggesting that nonhuman voices exist, but lack this particular communicative quality. His contention seems to be that the human voice touches the soul because it somehow already carries or reflects the soul of the one producing its sound. It reflects a "mental state" that shows the emotive range and evolutionary status of the speaker.

The idea of a soulful, communicative voice as a point of distinction between humans and animals was an important element of Darwinian understandings of humanity, but goes back to a much older debate. As numerous nineteenth-century commentators were clearly aware, the discourse of a voice that distinguishes humans from animals can be traced back to the ancient Greeks.[4] The thought seems to have lost urgency in Europe in the Middle Ages, as the vocal power to move was connected to divinity in greater measure than to humanity. But in the late eighteenth century, the definition of humanity and the voice came under relentless scrutiny, and their relation became, again, an object of systematic intellectual thought and scientific research. Debates on the status of the human voice and speech soared as religious understandings of the word were slowly backgrounded and the soul lost legitimacy as the defining characteristic that drew a line between innocent beasts and moral men. Plato's statements on the ability of "Saturn's children" to speak to animals were reexamined.[5] Yet, the association between humanity and the right to speech remained intact. Rather than as a universal mechanism, the voice gained a place as a means to articulate highly personal character traits (e.g., by expressing emotions) and to be recognizable as an individual. Modern researchers on the voice did not really espouse

Socratic or Platonic ideas on the voice, but they were happy to refer to the ancient philosophers' interest in their chosen subject, which conferred some intellectual gravity to their otherwise new and seemingly frivolous field of study. They were also quite keen to quote Plato's succinct phrasing of the connection between the human capacity to speak and individuality: "Speak, that I may know you."[6]

The idea that the voice on the one hand reflects something of its owner's soul and on the other hand touches upon the soul of its hearers is not foreign to us now either. The spoken word still feels more personal than the written one. We still occasionally talk straight from the heart, and are touched by the sincere or tender sound of a fellow human's voice. The sounds of our voices, or our capacity to organize those sounds into language, still serves as a shorthand for humanity, ensuring one's readability as a human and as an individual. In fact, the individuality and recognizability of our voices is such that we treat someone's voice (on a recording, for example) as a stand-in for that person as a whole, and hearing the voice of the deceased sometimes seems to temporarily bring them back to life. Our voices announce our presence when we enter a room, reveal our identity when we pick up the phone, and confuse us when we hear them played back to us.[7]

Because we identify so strongly with the sound of our own voice—which seems to emerge so intuitively from our throat—it is easy to imagine it as natural, given, and eternal. Even singers, who have consciously cultivated and developed their sound, tend to attach enormous importance to the naturalness of their vocal performances, and invest them with their personality.[8] Yet it is in professional singers that we also glimpse the historicity of the voice: generations of recorded singing allow us to hear unequivocally that the (desired) sound of the human voice has changed even in this short period of time. The divas of today do not sound like Maria Callas, and Enrico Caruso's voice is inimitable partly because it no longer represents the current tenor voice.

This book aims to show that changes in these sounds go beyond evolutions in artistic taste or fashion, and that the historical evolution of vocal sound also reveals social changes and their articulation. It sets out to demonstrate that the very idea of a highly personal and identifiable voice is itself inherently modern. This idea was born from scientific endeavors to map the larynx and to observe its movements. It derived from the cultural work of distinguishing between voices and the attempts to legitimate the lines of demarcation between them through studies of anatomy and measurements of sound. It depended on the corporeal work of enlarging lungs, training tongues, and disciplining jaw muscles. And it was connected to the growing insistence on the importance of the self-governing individual for a society ruled to a great extent by the spoken word. In the nineteenth century, democratic representation was spread and expanded to include ever larger groups of citizens. The history of the human voice in the nineteenth century is to a great extent one of fluid imbalances of power, and of shifting lines of demarcation: between sounds heard as voice (or: legitimate speech) and those heard as noise or shrieks; between those that articulate genuine emotions and those that merely resound with instinct; between those allowed to be heard publicly and those that were not; and between humans and animals.[9]

A Century of Vocal Materiality

I will set out to sketch these shifts for, roughly, the whole nineteenth century. The story begins with a surge of interest in the physical characteristics of human and animal voices—their mechanics, so to speak. In 1769, Wolfgang von Kempelen had attempted, reasonably successfully, to mechanically reproduce the vocal organs. His "Speaking Machine" did not quite speak, but presented a convincing imitation of the human voice nonetheless—one that was right at home among the numerous android automata of the time:[10] it spoke eloquently of human invention and industry, and allowed for fascinating discussions of humanity in its artful imitation of a central feature of man. Ultimately, however, it referred only obliquely to the present, living, breathing, feeling human body. Around the same time, voice pedagogy gained institutional stability as well: in 1795 the Conservatoire de Paris was founded. The Royal Academy of Music in London followed in 1822, and the Conservatorium der Musik zu Leipzig became, in 1847, the first college of music in Germany. The beginning of the nineteenth century presents a period in which fundamentally new ways of thinking about gathering and spreading knowledge about the voice, as well as new strategies for its cultivation, were developed in scientific and artistic circles. Throughout the century, this knowledge was broadened, subjected to increasingly systematic analysis, and integrated in education. Roughly in the middle of this period, around 1855, an innovative invention was introduced that would change the field and its methods of inquiry: the laryngoscope. A handful of physicians and canto teachers developed, almost simultaneously, a system using mirrors in combination with natural and artificial light sources to observe the vocal folds. The laryngoscope allowed researchers to see the voice, as it were, and observe it as a living thing. It would have an enormous impact not only on anatomical studies and medical research, but also on the vocal practices of singers and speakers. Teachers of canto were as interested in the instrument as surgeons, physiologists, and speech therapists.

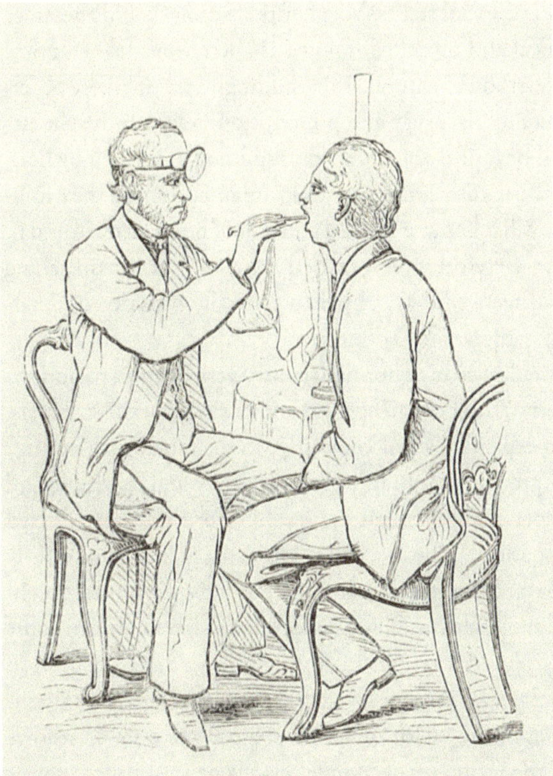

Figure 1. George Johnson, *The Laryngoscope: Directions for its use, and practical illustrations of its value in the diagnosis and treatment of diseases of the throat and nose. Two lectures delivered at the royal college of physicians*, Robert Hardwicke, 1864. Courtesy of the Wellcome Collection.

The instrument was embraced widely at the time of its invention, exerting its influence on a range of professional arenas. Historiographically, it symbolizes a range of broad cultural changes concerning the observation, cultivation, and meaning of the modern human voice. The explicit notions of interdisciplinarity embedded in the laryngoscope, which made it obvious that physiological knowledge could be of benefit to artists, were indicative of more general changes in the developing field of laryngology. They resonated with a wider move toward a more scientific approach to vocal practice on the one hand, and an increasingly somatic practice of vocal science and medicine on the other hand. They also reflected a perhaps more abstract but no less influential sociocultural phenomenon. Laryngoscopists' insistence on combining (medical) scientific knowledge and artistic craft also contributed to a more general conflation of vocal health with esthetics. In other words, in conflating beauty with health, experts in the nineteenth century started to consider vocal practices as reflections of the vocalizer's inner state and social status. Vocal imperfections could therefore be read as signals of physical and moral pathology, as aberrations from a norm defined by social laws understood as natural. Their findings were not mere observations, but an odd mixture of discovery and prescription in which the characteristics of the healthy and beautiful voice were presupposed, informed by both a long history of Italian craftsmanship and new Darwinian notions of the evolutionarily superior human. What became normal and would be termed natural in the nineteenth century was therefore closely related to a tradition of painstaking cultivation carried out by singers who were by then generally considered unnatural (and dehumanized) creatures: the castratos.

The construction and cultivation of normality and its connection to notions of humanity are at the heart of this book. The ideas and practices of the scientists, educators, and artists who expanded vocal knowledge in the nineteenth century reflect wider notions and practices of middle-class propriety and its visible and audible articulation. The material on which the research for this book is based (scientific treatises, therapeutic handbooks, and educational manuals) was easily available at the time of its publication, but has since mostly sunk into obscurity. The few scholars and musicians cited in this story who have retained some level of fame have mostly played a role in institutional histories of education and science. In other words, they have been read as thought leaders, as examples of normative discourses about health and musical quality or taste. In studies that feature, for example, the influence of Manuel Garcia or Paul Broca, power is generally ascribed to the institutions or authors producing aesthetic norms or scientific standards. Their theories of language production became the norm, or their methods for vocal education were adopted internationally. Rather than these explicitly articulated standards themselves, however, I am more interested in how normative publications also reflected more mundane historical behavior. Or, more specifically, in how nineteenth-century experts distinguished between the different vocal behaviors they observed, how they ascribed different expectations to different speakers and singers, and—perhaps most importantly—how different vocal practices were consciously developed, cultivated, and curated for different social contexts. In other words, in this book, the relation of various singers and speakers to power and its vocal articulation is at stake.

The conflation of (corporeal) esthetics with morality and ideas of naturalness and health was hardly limited to the sound of the voice. Medical historians have shown that numerous visible pathologies have been read as signs of internal disorder, and the connection between middle-class conventions and perceived naturalness has been demonstrated in numerous contexts.[11] The voice and its use present yet another site where this connection was tightened throughout the nineteenth century: various vocal exercises, practices, and sounds that were defined as healthy or beautiful served to articulate particular middle-class social and behavioral repertoires. However, in focusing on what is necessarily a repetitive practice of vocalization,[12] a history of the sounding modern voice also offers a new perspective on how nineteenth-century middle-class citizens delineated their own identity from that of others whose voices they considered worthy of observation.

Tracing changes in the way knowledge about the voice was accumulated and expressed, and—especially—in the social meanings attached to manifestations of the voice, gives us a particular insight into the slow evolutions in the practices of demarcation that characterized the performance of new modern identities. Although the voice is issued from a body, and can almost intuitively be linked to its owner's physicality, it represents one of the few corporeal characteristics of humans that can be instantly manipulated and consciously controlled. Changes in the voice and its sound are difficult to trace because of the voice's performative and transient nature, but those same characteristics have also enabled the voice to change more consciously (and perhaps more quickly) than other facets of the body.[13] Creating a proper face or body might seem well-nigh impossible, but proper vocal delivery—like posture or gait—appears controllable.[14] Despite the returning trope of the individuality and naturalness of the voice, the treatises and manuals under scrutiny here show quite clearly that the proper voice was a matter of continuous and largely consciously performed work. Considered separately, singing manuals and scientific treatises on vocal health demonstrate aspects of this cultural work: they show the development of particular exceptional voices (exceptionally beautiful ones, or exceptionally unfortunate ones).

Taken together, however, the impressive stream of publications by experts across such disciplinary divides also reveals something about the social salience of vocal health and beauty. It is significant that the difficult cohabitation of seemingly given corporeal features enabling vocalization and the voice's calculated cultivation was a matter of concern in both the medical (or scientific) and the artistic worlds. Throughout the nineteenth century, physiologists, surgeons, therapists, educators, and singers exchanged knowledge, cited one another's work, and collaborated on publications, but also attacked one another's work with passionate abandon and systematic precision.[15] Laryngoscopists were not alone in insisting on the necessity of combining physical experiments and musical tradition, but the conviction that artists and scientists benefited from each other's work did not ensure a continuously harmonious relationship between both fields. The relationship between both sites of knowledge, and the means and styles of exchange (e.g., in how they represented each other's findings and relevance) changed considerably over time. A study of these different expert discourses is necessarily limited to a fragmented view of nineteenth-century vocal culture and its ensuing practices, but combining the medical and musical world does allow for a particular inquiry into the conflation of propriety, health, and beauty, and the cultural work involved in producing it that is at stake here.

Quite a lot of that cultural work was invested with notions of differentiation or demarcation. First, experts were involved in the "boundary work" that defined their own fields of expertise and their own identity as bearers of knowledge on the voice.[16] Perhaps because the fields of laryngology and speech therapy were so new, they were populated by people with wildly divergent and often questionable or unconventional credentials. All self-appointed experts on voice, but especially those longing to establish themselves as medical or artistic authorities, were engaged in a constant quest for (institutional) legitimacy. Second, differential notions of the proper voice were forwarded—and changed—throughout the century as experts developed sophisticated connections between vocal practice and social status (often by legitimating or cultivating existing cultural expectations). And thus women's voices were distinguished from men's, men's from children's, English from Italian voices, civilized from savage ones, and so on. As demarcations were drawn and redrawn between these different voice types, both artists and scientists struggled to marry developing notions of the observable natural human voice and the practice of articulating specific identities (identities also subject to considerable change). What was a healthy, attractive, or proper voice in one body could be profoundly jarring in another one. Singing manuals and scientific treatises alike abound with examples of such misplaced voices. Female tenors, strapping young men with light falsettos, or serious politicians with frivolous piping or husky voices were as fascinating as they were disturbing in a world dominated by clear vocal norms and expectations. The enthusiasm of the first half of the century for vocal mechanics, cemented by notions of an easily distinguished human voice box, made way for a more holistic approach in which not only was the whole body involved, but its social contexts and culturally charged movements were taken into account as well. Rather than simply peering into a singer's or orator's throat, physicians, therapists, and theater audiences were taught to associate different body types with types of larynxes and the sound they produced.[17] The story of this changing landscape of identities articulated, quite literally, by voices that were both consciously developed and minutely observed, is relevant beyond the confines of the laboratory, the hospital, the stage, or the music classroom. Although this book owes much of its empirical foundation to the discourse of experts, its goal is not to reconstruct historical norms—nor does it seek to find out what ideal or normative voices of the nineteenth century "sounded like." In aiming to provide a social history of the modern voice, I tell the story of how various practices of speaking, stammering, singing, shouting, and other modes of vocalization were implicated in "doing" sociability and social distinction in modern Europe. As various social historians have pointed out, practices like "visiting," informal gatherings in both domestic and public settings, and "meetings" were central aspects of nineteenth-century social life, particularly for the middle classes.[18] In all of these modes of sociability, vocal practices (both speech and song) played a central role, which has largely been left unexplored in historical studies.

The structure of this book more or less follows the development of different expectations and practices of the voice "at work," focusing in turn on the boundary work among experts and the cultural repertoires their work expressed and delineated. The invention of the laryngoscope appears as a turning point in this story. It turns laryngologists' and physiologists' attention to the living voice (which brings them closer to the world of voice cultivation and artistry) and thereby penetrates the relatively insular world of great vocal traditions and their maestros. Importantly, it also rang in a period in which the raw matter of the vocal organs became inextricably connected to the use their owners made of them. The introduction of the laryngoscope almost completely eradicated the research on dead larynxes that had been carried out before, and coincided with the decline of surgery as a cure for speech impediments.[19] It changed the definition of the voice profoundly as it located its existence in the practice of manipulating the organs rather than in their mere material presence in the throat.

The focus on expert knowledge and its accoutrements, such as the laryngoscope, means that the number of people involved in both the production and consumption of the discourses that are examined in this book is relatively small. The impact of these experts' works on vocal practice on a wider scale is difficult to gauge: manuals for teachers are likely to reach beyond their actual readers, and medical manuals indirectly affect patients as well. However, the most important argument for these prescriptive, normative, and expert-driven texts derives from the constant interaction between practices of expertise and cultural vocal practice. On the one hand, the questions and methods forwarded by experts derive from preoccupations that carry legitimacy within their cultural and historical context: the reoccurring attempts to unravel the links and tensions between vocalization and humanity, for example, resonate with a wider cultural preoccupation with the limits of man in a time when Darwinism and colonialism are changing that concept's Enlightened definitions. On the other hand, as prescriptive as these works are, they do contain descriptions of voices as well (e.g., in the form of case studies of individual patients' voices). Moreover, the very prescriptive nature of most texts connects them to changing practices as well: the voices projected in therapeutic and artistic manuals may have been slightly idealistic representations of voice, but it is reasonable to assume that readers would indeed aspire to these ideals and adopt at least some of the practices prescribed in a manual they had taken the trouble to purchase (or were forced to assimilate in order to gain a degree).

In the first part of this book, the accumulation of expert knowledge and its prescription to singers and speakers in the first half of the nineteenth century is examined. Chapter 1 sketches the emergence of a medical and scientific field of inquiry, characterized by territorial disputes between surgeons and physicians on the one hand and speech therapists, elocutionists, and singers on the other hand. Through the writing and rewriting of their own discipline's history, experts laid claim to authoritative knowledge, delineated their subject, denounced others as quacks, and effectively created a field of study in the first half of the century. The second chapter discusses how the findings of these experts in emerging fields were translated into prescriptive discourses on vocal practice, and focuses on the tensions between attempts to map out the vocal organ as an anatomical given, and the simultaneous insistence on moral and physical labor that would ensure a proper development of the voice—especially for children.

Part two looks at how the field of knowledge and its social context changed after the introduction of the laryngoscope in the second half of the nineteenth century. In Chapter 3, the increasing cooperation of experts from different disciplines is examined: as the living voice, rather than the isolated larynx, came under scrutiny, the body as a whole became the object of study for both scientists and artists. Moreover, aside from the generic (or perhaps universal) body of the anatomical model, the very real bodies of authors and their audiences played their role in the discourse: this is also the period in which self-help manuals were gaining popularity, through which stammerers, singers, and others increasingly attempted to seize control over their own voices by building their own expertise. Chapter 4 looks at the increasingly important idea of the voice's controlling mechanism: the brain.[20] In the second half of the nineteenth century, explanations focusing on isolated parts of the vocal apparatus met with mounting criticism. Stammering, for example, could no longer be explained by faulty musculature in the tongue, and the cutting of the frenulum was increasingly frowned upon. Likewise, the width and length of the vocal cords were no longer measured with the same obsessive attention, since they could not account for voice quality or timbre. The attention of experts not only shifted to the body as a whole and its willful direction; it also centered on the role of the will in speaking. As a corollary to that, attention shifted from the voice (as a human attribute) to "speech" (the explicitly human way of controlling and using the voice).

Interspersed between the chapters are brief "intermezzi," highlighting particular characters and objects from the history of modern vocal science and education. Where the chapters are, necessarily, somewhat synthetic in their approach, seeking out the connections and shared discourses and practices of science, music, and education, these intermezzi show how these widely shared approaches were carried out in practice, often quite idiosyncratically, by particular actors in the field situated in particular corners of the artistic-scientific ecosystem, such as one of the authoritative women's voices in vocal education, a stammering professor, and a handful of monkeys.

The fifth chapter, finally, sketches the voice's entry into the twentieth century, and the introduction of audio recordings.[21] The invention of the phonograph would have undeniable consequences for singers, whose work could now be spread such that their unique voices would enter their audience's living rooms. In physiological, psychological, and neurological research, the application of the phonograph was less triumphant: audibly reproduced sounds were hard to systematize, and researchers struggled with notions of objectivity and distance as they tried to interpret the results garnered by a host of new audio machinery (it would play an important role in the growing field of spiritualism, however).[22] It seems that, after a century of attempting to grasp the voice and anchor it inside the vocalist's body, the introduction of audio recording made the voice fleeting and eerie again, linking it to notions of individuality and humanity that resonated with earlier notions of the soul. For an early twentieth-century audience, Enrico Caruso *was* his voice: a sound that was deeply moving, uniquely connected to this one star of the stage (and therefore highly individual, modern, and recognizably Italian), but that could be replicated without the interference of the singer's actual body.[23] The introduction of recordings and radio granted the spoken and sung voice new authority and reach, as it marked the end of an age of vocal materiality.[24]

History Listening In

The history of the connection between the voice and the body is something of a historiographical blind spot. The perceived ahistoricity of the voice is part of the issue. Producing voice, in either speech or song, comes "naturally" to most people, and that obscures its cultural and acquired nature. However, more important than that has been the isolation of the voice in defined areas of cultural inquiry. When the historical voice comes under scrutiny, it is typically in connection either to music or to language. It seems that, as soon as the voice leaves its owner's throat, it is catapulted into a space of either cultural politics or artistry, and its production is entirely forgotten.[25] This is not to say that studies on the singing voice or on the voice in speech have paid no attention to the owners of voices and their social and cultural contexts. John Potter's work on the authority of style, for example, focuses very specifically on differentiation between singers, and the historical and spatial embeddedness of the singing voice.[26] His more recent history of singing, with Neil Sorrell, is similarly interested in the geographical and cultural embeddedness of different singing voices and their meanings.[27] The anthropological perspective employed in Potter and Sorrell's history of singing gives considerable attention to who the singers are, and how that affects the nature, meaning, and impact of their voices. As a 2015 forum in the *Journal of the American Musicological Society* suggests, studying voice is culturally salient, pushing us to ask why the voice and its effects are of importance "now."[28] Additionally, James Davies's reflections on the voices of Henriette Sontag, Maria Malibran, and Jean de Reszke demonstrate that such questions on the cultural inflection of professional singers' voices go far deeper than matters of style: historically contingent notions of breath, body, and culture are deeply implicated in how singers carried out their craft.[29]

Likewise, Arlette Farge's history of French spoken voices in the eighteenth century is carefully attuned to the connections between location, social class, and pronunciation. Her study of "the voice of the most vulnerable" (Farge shows particular attention to the voices of children, women, and "the people") shows that modes of speaking and political legitimacy were intimately intertwined in prerevolutionary France, not in the least through the differentiation between French and *patois*.[30] In her introduction, Farge notes that in the wake of the period she describes, the voice gained "a primordial place in literature, as it does for certain historians," and this seems to have been true in Britain as well. Ivan Kreilkamp's *Voice and the Victorian Storyteller* argues convincingly not only that the voice "persists in the discourse of print culture" throughout the nineteenth century, but that Victorian writing—and the novel in particular—brought the storyteller and his voice into being.[31] Yet, while these accounts of the voice as musical or linguistic history illuminate the processes of differentiation that interest me here, they have surprisingly little to say on the production of difference through vocality. Their attention to the sound of the historical voice is almost exclusively turned toward reception, that is, what could be heard by the contemporary audience. The so-called period ear (a largely discursive reconstruction of the sociocultural background informing the audience's perception) is central in these studies, in which the somatic perception of voice and ambient sound is at issue—or, in terms of communication theory, in which the receiver rather than the sender plays a central role. The "ear" thus focuses on the voice as a bearer of discourse—whether by performing language or a musical and esthetic repertoire. In 2000 Steven Connor stated that "if [a history of the voice] were to be undertaken, it would have to be in terms of the struggle between legitimate voice and the different forms of spacing, sensing, and forcing embodied in the ventriloquial voice."[32] The handful of histories concerned with the voice have not managed to address these issues of spatially and somatically defined corporeality yet.[33]

The body does have a habit of sneaking into analyses of the voice, though, even in stories concerned with art or power. A number of studies dealing with the voice on stage have focused on the issue of gender—thus (indirectly) dealing with the intersection of the social and the somatic, and the role of the voice in expressing what is usually understood as a corporeal category. The most obvious historical example of artistic voices articulating corporeal categories is perhaps the castratos: unique voices emanating from mutilated male bodies. The result of far-reaching artifice, the castrato voice not only articulated the traumatic manipulation of gendered physicality, effectuated through surgery, it also echoed long years of training and an even longer tradition of vocal education and conventions of the stage. Martha Feldman's reflective study of this voice shows how understandings of sacrifice contributed to a counterintuitive construction of ideal masculinity in the castrato's body and voice.[34] The golden age of the castrato preceded the period of obsessive naturalization of the voice that I am concerned with here. But the memory of their inimitable voices lingered, and modern imaginations of the castrato sound informed many of the theories surrounding the gendered physicality of voice. Moreover, as Naomi André writes in *Voicing Gender*, the disappearance of the castrato from the stage, and the simultaneous rise of the "second woman" in opera at the beginning of the nineteenth century, initiated substantial changes in the associations between pitch, timbre, and gender on the operatic stage.[35] The extreme focus on pitch, embodied by the "Angel's cry" on stage, was likely absent outside the world of professional musicianship—regardless of whether said angel was an unconventionally gendered man or an equally ambiguously embodied diva.[36] Yet, the connection between gendered voice and embodied authority has been traced for the speaking voice as well. Christina Luckyj, focusing on early modern theatrical texts, goes further back and demonstrates the gendered notions associated with silence. She thus offers important insights on the history of the female voice by suggesting that silence might "offer a rival, less highly regulated space" than speech, which "bears traces not of personal agency but of institutional constraint," so that "the men who can speak" were not necessarily "freer than the women who keep silent."[37] Gina Bloom's analysis of squeaky and other voices adds the dimension of age to the gendered voice, as "the presence of cracking voices on the stage had the potential to be not only an aesthetic but also an ideological liability; for the cracked voice, as it evinces a body in transition, reveals the precarious, shifting nature of male identity."[38] Just how changeable the male voice and the identity it constructed could be is obvious in Ann-Christine Mecke's history of the voice break in the German choral tradition: in the eighteenth and nineteenth century, the meaning and use of the pubescent male voice changed considerably, and was increasingly pathologized.[39]

Whereas in the eighteenth century "physiological development and tessitura" were only "loosely connected," later notions of vocal nature and normality demanded a strict separation between male/female and children's/adults' voices, in which the mutating voice could barely find a legitimate place.[40] Creating room to understand those voices that did not conform to such strictly defined categories of binary gender requires a "queering [of] the pitch," as Philip Brett has put it,[41] including subjectivities that often exceed the boundaries of what is commonly understood as bel canto, classical music, or simply good taste. Musicologists concerned with contemporary popular music have contributed significantly to this new perspective,[42] and in this book I turn mostly to Freya Jarman-Ivens's work *Queer Voices*—which features Maria Callas's voice next to those of Karen Carpenter and Diamanda Galás—to delve into the role of vocality and music in the performativity of gender and sexuality itself.[43]

These works on the musical flaw and vocal cracks go a long way to explain how voices can be material manifestations of gendered identities, both now and in the past, making the historicity of the voice tantalizingly intangible. Philosophical explorations of gender and sexuality and the voice—most notably Adriana Cavarero's *For More Than One Voice*—seem to suggest an almost eternal connection between the maternal and the singing voice.[44] Historical examinations such as Luckyj's or Felicia Miller-Frank's study of the process of associating femininity with the singing voice and the translation of that connection into psychoanalytical categories in the nineteenth century contradict some of these reflections.[45] However, it remains unclear how the association of vocality and (gendered) physicality that seems to underlie some of Cavarero's analysis has come into being. Evolutionary, hygienic, and physiological insights developed apace in the nineteenth century, and the impact of these incarnations of modernity on the nature and meaning of vocality largely remains to be studied. Moreover, the period was also marked by advances in psychoanalysis, a discipline that has continued to exert influence on the study of the voice, including in the current field of voice studies through the work of Mladen Dolar.[46]

One way into a more finely tuned history of the voice and its corporeal production could be provided by an examination of the role of the voice in medical history. Denyse Rockey's history of speech pathology in nineteenth-century Britain shows that this is—particularly for the nineteenth century—a very rich terrain. In 1985, Rockey wrote in her introduction to *Speech Disorder in Nineteenth-Century Britain* that "history is a new frontier in speech pathology."[47] In her study of stammering cures and their institutionalization in the nineteenth century, she confronts that frontier, suggesting various avenues for further historical exploration. Rockey's work illuminates important epistemic changes in the way speech pathology was approached and understood, and thereby also throws light on some of the social and cultural consequences of the changing definitions of "normal" and "impeded" speech. Although her work does not speak directly on the sound of the voice, it does demonstrate the intimate connection nineteenth-century observers assumed between the body (or at least the throat) and the spoken word. It also shows how sources detailing the accumulation of knowledge on speech can inform an analysis of its sociocultural meaning.

More recently, this suggestion has been further developed by scholars in the emerging field of dysfluency studies: Joshua St. Pierre's *Cheap Talk*, for example, delves into the constitutive and often pernicious role of the discipline of speech pathology in co-constructing fluent speech and the ideal of the productive employee of modern industry and neoliberal societies.[48] The longer history of speech therapy and stammering cures plays a significant role in this book as well, due to its important place in the production of a standard for healthy speech in the nineteenth century—a standard that, as Riley McGuire, Daniel Martin, and others have shown, was closely connected to simultaneously developing ideologies surrounding maturity, heteronormativity, and civilization.[49] George Catlin's ideas on the naturally fluent Indian, for example, were part of the same cultural edifice that would also construct the so-called Black voice. Jennifer Stoever's study on what she has termed a "sonic color line" discusses how racialized voices were heard as distinctly (and discriminatorily) different well into the twentieth (and arguably into the twenty-first) century.[50] That race was heard and judged in both individual and collective voices throughout modernity was perfectly obvious, not least to those possessing an ethnically marked voice. In both rhetorical performances like those of Olaudah Equino, and literary works such as W. E. B. Du Bois's *The Souls of Black Folk* or Toni Morrison's *Song of Solomon*, recognizably Black voices take center stage.[51] Contrary to Du Bois's and Morrison's fluid representations of various Black voices of infinite modulation and flexibility, however, most representations of racialized voices throughout the nineteenth and twentieth centuries imposed inflexible categories on the timbres and rhythms of race—or, as Nina Eidsheim has put it, "the racialization of vocal timbre exemplifies . . . processes of ossification."[52] In large part, this ossification is the result of the particular modes of knowledge production and authority claimed by those who could legitimize their expertise on vocal health, beauty, and classification, and whose works are now available to us as sources for the history of vocal norms and practices.[53]

James Stark's history of bel canto similarly relies on processes of the production and spread of expert knowledge to study the growing intertwinement of science and art in singing methods (more or less from the nineteenth century onward, with great attention to Garcia's research and teaching). His interest in the matter is highly practical: focusing on a number of vocal problems, bel canto provides a historical context for the physical and technical knowledge of contemporary singers. Writing about registers, for example, he provides a detailed comparison between the nineteenth-century three-register theory and the older two-register theory, drawing on the work of Garcia, Tosi, Mancini, and a host of other authors. It is a sophisticated analysis of the changing views on vocal mechanics, but does not leave much room for the social associations and implications of vocal registers and their meanings beyond the stage.[54]

Studies on the particular capacity to vocally express (socially differentiated) notions of propriety, it seems, have largely been limited to the terrain where voice meets language. Studies of parliamentary speech, such as Christopher Reid's account of the eighteenth-century House of Commons, Ben Griffin's analysis of the embodiment of masculine identities in the nineteenth-century House, and Joseph Meisel's more broadly conceived study of the practices and discourses of public speech in nineteenth-century Britain all go beyond notions of language and style. In all these considerations of public and professional speech, it is obvious that the practices of political eloquence were closely interconnected with social and corporeal stylizations of self.[55] How a speaker positioned himself, socially among his peers and physically at the pulpit or on the benches, clearly influenced the impact his words could have. In different ways, and from different perspectives, Reid, Meisel, and Griffin all stress the historical importance of delivery in political oratory. Their interest extends to various educational practices, the spatial arrangements of the House, and even facial grooming and expressions, but leaves the actual act of speaking, as a physical and acoustic event, largely unquestioned. Studies that do center speech (or its absence) as such, like Tom Wright's research on the history of lecturing, and Henk te Velde's and Theo Jung's work on, respectively, parliamentary speech and silence, have pointed out the close intertwinement of political speech and representation or political influence and presence.[56] Likewise, Ingrid Sykes's examination of the "voice of the blind" in nineteenth-century Paris is attuned to the somatic, but largely approaches voice as a metaphor for public presence, emancipation, and political participation.[57] This is indicative of a more general tendency to use the language of voice as a synonym for making oneself heard. The tradition of associating the voice with agency and power is as much part of the construction of the modern voice as its materialization and embodiment in the nineteenth century.[58]

The impossibility of really distinguishing between the history of speech and that of the human voice thus derives from more than the simple fact that the voice is (almost exclusively) produced simultaneously with the production of discourse—either in speech or in song.[59] It also depends on the saliency of the voice as a metaphor for political activity as it is mobilized in ideas on modern democracy.[60] The question of vocal propriety was therefore not limited to notions of beauty and health, but also included notions of political belonging—as Katherine Bergeron has argued in her study of the conflation of language, pronunciation, and melody in the French Belle Epoque. By the end of the nineteenth century, "speech had turned into a matter of national interest, the spoken word had come into clearer focus, and so *clarté* now became linked to the actual instrument of French, the voice."[61] Picking up more or less where Farge leaves off historically, Bergeron traces the difficult process of the creation of French as a standardized spoken language, and connects its evolution to the simultaneous development of a particularly French adaptation of the *Lied* (exemplified by composers such as Gabriel Fauré and singers like Charles Panzéra). In Paris around the turn of the century, linguists, physiologists, and musicians appear to have contributed to this process of linguistic standardization and musical eloquence quite consciously, each from their own discipline but often converging at the Conservatoire, the Sorbonne, or the Abbé Pierre-Jean Rousselot's laboratory for phonetic research at the Institut Catholique.

Bergeron's musicological analysis of cross-disciplinary practices of gathering and spreading knowledge on the French voice of the Belle Epoque points to important lacunae in historical research on the voice. It shows, first, that questions of language and its political currency cannot be addressed without taking into account the very mundane matter of uttering words, and how expectations surrounding that practice changed. Most studies concerned with speech veer into the discursive or linguistic once the thorny issues of meaning and social impact emerge, but the late nineteenth-century observers in Bergeron's work clearly presumed that there existed numerous links between vocalization and the social or the meaningful.

Additionally, the seeming isolation of the Belle Époque in vocal matters shows just how little work has been done to put this new view of the French language and its sound into perspective: both the period preceding the Belle Epoque and the international context of the very French *bon usage* are unclearly documented for the most part.[62] We know very little about what nineteenth-century vocal utterances sounded like to their audiences. As literary scholars like Liesl Yamaguchi and Tanvi Solanki have suggested, part of the answer might be found at the interface between literature and science: the nineteenth century represents a particular moment in which poetry, linguistics, physics, and performance converged in new ways, while strong claims of disciplinary divergence emerged at the same time.[63] The goal of this book is, therefore, to approach the nineteenth century as the heir of what has been discovered by Bloom, Farge, Luckyj, and others on the early modern voice, but also as the cradle of the modern voice that would further develop in the Belle Epoque and go on to thrive into the age of recording. Focusing on the corporeality of the voice, I am mainly concerned with the vocal production of sound, the conventions and rules guiding that production, and the wider meaning attached to practices of vocalization. This is a book about speaking, singing, and stuttering rather than hearing, and aims to contribute to a burgeoning field of sound history—in which the voice has thus far played a modest part.[64]

Reading with the Grain

The field of sound studies has informed most of the methodological decisions taken in this book, and partly defines its perspective. Sound studies is—as defined by some of its proponents—an "interdisciplinary area that studies the material production and consumption of music, sound, noise and silence and how these have changed throughout history and within different societies,"[65] and this study is in keeping with that definition. It focuses on the material manifestation of a particular way of producing sound, is specifically concerned with (historical) change, and strives for a balanced relation between historical and musicological approaches. Historical research within sound studies has been particularly effective at untangling the links and tensions between technological advances in acoustic recording and the development of listening as a range of embodied and encultured practices.[66] Despite its low-tech nature, a history of the voice may not be so dissimilar to the history of the gramophone or the telephone. In *The Audible Past*, Jonathan Sterne stated that "prior to the twentieth century, very little of the sonic past was physically preserved for historical analysis at a later date. So it makes sense to look instead at a particular domain of practice associated with sound. The paper trail left by sound-reproduction technologies provides one useful starting point for a history of sound."[67] The anatomical drawings, self-help therapies, and singing manuals that are at the basis of this study represent, arguably, the paper trail left by sound production technologies of a particularly "vibrant" kind. The human larynx appears in these works as a tangible producer of sound. And like other sounding instruments of the time, its definition and use were never innocent either. Larynxes and the vocal organs were "shot through with the tensions, tendencies and currents of the culture from which they emerged." Much in the way sounding devices carried the marks of human ambition, the "wishes that people grafted unto sound-reproduction technologies" that became "programs for innovation and use" were grafted onto the organs of speech as well.[68] The history of voice and vocalization, finally, is like the history of sound technology in that it can be read as a history of modernization: the accumulation of knowledge about the voice, its systematization, and its resulting practices of vocal identity and individuality "represented"—like the sound reproduction technologies in Sterne's story—"the promise of science, rationality, and industry and the power of the white man to co-opt and supersede domains of life that were previously considered to be magical. For their early users, sound technologies were—in a word—modern."[69]

If the modern voice can be understood as a means for the production of human sound, part of its paper trail can be read as a record of the construction of its constituent parts. This does not enable us to reconstitute the historical voice in the same way one could reconstruct historical technology or musical instruments in an effort to recover its sounds. But the affinity between the meticulously drawn manuals for mechanical sounding devices and the phonographic claims of nineteenth-century voice experts (epitomized at the end of the century by acoustic recordings) can alert us to the qualities of medical and educational texts of this period as recordings of the time. In a sense, of course, all language serves as a way to record the voice in speech.[70] But I would argue that in the nineteenth century, with its almost unwavering belief in systematization, its taste for scientific discovery and observation, and its very peculiar understanding of how voice could be written, the documents used here can be read as recordings in a more meaningful and a more corporeally anchored way. A recording is, according to Jonathan Rée, "a technique for generating copies of an original, rather than maintaining it in existence."[71] Drawings, texts, sheet music, and so on do not reproduce the nineteenth-century voice, but they do represent what contemporary observers considered to be a copy. They may not offer us the elements of the voice we would expect, but do give us an insight into which characteristics of the voice were thought to be essential at the time of recording. If the voice is in the first place a production of the larynx, it makes sense to picture it anatomically. If pitch is an essential feature of vocalization, musical notation will appear as an efficient way of representing it. Throughout the nineteenth century, those keen to document (vocal) sounds came up with various novel methods to do so. The nineteenth century saw the invention of short-lived fads such as visible speech, photography of the voice, and the tonograph, but also of more durable methods of transcription such as shorthand (originally often called phonography), which was presented as a way to somehow capture speech as it was being produced, rhythmically following its patterns in a way ordinary writing could not.[72]

Approaching these graphic representations as if they are recordings seems counterintuitive. Especially in the case of musical scores, our idea that their "sounded writing" represents an almost Platonic idea of "the work" that is produced by a composer and then reproduced by performers[73] makes it hard to see them as copies. Surely the composer's ideas are the original, and multiple performances its copies? In musicological analysis, attuned to the performative nature of music, this idea of compositorial activity has been convincingly analyzed as an instance of voice in a more abstract sense. Carolyn Abbate's attempt to "restore to music criticism a sense for the physical force of music" by calling attention to opera's "unsung voices" is an appealing one. "Music," Abbate claims, "like the human speaking voice has the capacity to assail us with its sheer sound." What Abbate is restoring is, in a sense, the capacity of voice to perform presence, to embody subjectivity even when divorced from an individual voicing body. In music, indeed, the voice can be both "voice-object" and "the source of sonority, as a presence or resonating intelligence."[74] This book takes a more pragmatic stance, however, and zooms in on the connections between human voices and modern individuality on a social, rather than a philosophical, level. It therefore draws on scores as documents of a social reality, rather than a representation of the composer's somatically informed voice.

I do not mean to suggest that scores or texts are necessarily direct consequences of a particular moment of vocal sound: they are not transcriptions of actual performances. What I do propose is to see all these recordings as the indirect heirs of audible performances, inspired by the somatic culture surrounding their production. "Recordings," then, are not only "not adequately understood as reproductions of performances," but they rather "represent performances through complex processes of cultural signification." Nicholas Cook's description of how we hear audio recordings of music provides an interesting starting point to think about these earlier non-acoustic recordings: "We hear recordings as performances, in other words diegetically, even when we know that the performance represented by the recording never took place."[75] Likewise, I want to approach scientific treatises, singing manuals, and visual representations of the voice as an integral part of the vocal performance practices of the past.[76] To their readers, representations and depictions of the voice were "recordings of performances," referring to a recognizable reality of performance, even if the particular performance present in the recording never took place as such.[77] In other words, based on these documents, readers could recall or imagine the sounds described, because the texts drew on a vast catalogue of acoustic memories, of familiar sounds.

The recordings I will analyze here present a heterogeneous collection of material. They are drawn from influential institutional collections in three cities: London, Paris, and Leipzig. More concretely, they are drawn from the Wellcome Library and the libraries of the Royal College of Music and Royal Academy of Music in London; the Bibliothèque Nationale de France (containing the historical library of the Conservatoire National de Paris), and the libraries of the Samuel Heinicke Institut and the Hochschule für Musik und Theater Felix Mendelssohn in Leipzig. While most of the material present in these collections was produced locally, the database mainly reflects the amalgamation of knowledge that was available to the interested reader in each of these cities—and they seem to have read enthusiastically across disciplinary and national boundaries. The personal collection of Henry Wood (held at the Royal Academy of Music) underscores this point: his bookcases held artistic and scientific work and—though overwhelmingly English—suggest at least some affinity with work produced abroad, both on the continent and in the United States. The corpus of sources here is thus an international and an interdisciplinary one, characterized by continuous exchange. Garcia may be cited as an example again, as he was active in both London and Paris. However, the inclusion of institutions from different disciplines and in different places also points toward several sites of distinction: scientists and artists may have read one another's work, but seem to have done so largely in different places. The project of standardized speech was high on the agenda in both Germany and France, but *Hochdeutsch* acquired other meanings than *le bon usage*.[78]

It is not my intention to consistently compare three national or cultural contexts, or two disciplines (which the material would not allow—Paris is not France, laryngology changed its face constantly, and Germany did not even exist as a singular nation for part of the period under scrutiny). However, what the choice of this handful of institutional collections does allow me to do is to illuminate the political, cultural, disciplinary, and linguistic locatedness of the modern voice. It shows that the voice is not only anchored in the modern body, but also influenced by the place from which it speaks. Much as Western-European ideas on the individual citizen and the vocal articulation of identity may resemble one another across borders (and the frequency with which experts quoted knowledge from a variety countries and disciplines certainly suggests some discursive unity), notions of vocal propriety were influenced by the political and institutional context in which they developed, and by the modes of knowledge accumulation within these different institutions. The interest in stammering, for example, arose at different times in different places: French experts dominated the field in the first half of the nineteenth century, after which English researchers rose to prominence. Changes in notions of legitimate research have therefore colored the definition and the social meaning of stammering in both countries—and vice versa.

What is ultimately at issue in my reading of these multiple recordings is akin to the "something more" Martha Feldman and Judith Zeitlin have playfully riposted to Mladen Dolar: the "paradoxical status of voice as something that hovers between embodiment and disembodiment [and so] raises the uncertainty of determining to whom a voice belongs."[79] Feldman and Zeitlin also conclude from this status a "need to situate specific voices within a specific historical time and place." While their own answer to that need guides them toward psychoanalytical approaches to materiality, mine leads me away from the *objet petit a*, and toward nineteenth-century understandings of the voice as a material and indeed physiological object. And, as Feldman and Zeitlin note as well, such a project almost unavoidably leads back to one of the foundational theories of voice: Roland Barthes's notion of the "grain." As will become clear throughout the chapters of this book, I have—like many other scholars of voice—reservations about some of Barthes's epistemological claims as well as his conclusions regarding vocal training and its effects, but continue to find value in "the support" his theory "lends embodied, identificatory practices of listening and practices of...

voicing that tend to afford agency to precarious and marginalized peoples."[80] And so, in many ways, the historical practices at the heart of this book are also understood to produce something like the *grain de la voix*: it is not divorced from language (according to Barthes it is "the materiality of the body speaking its mother tongue"), but privileges the "geno-song": "the volume of the singing and speaking voice, the space where significations germinate 'from within language and its very materiality,'" characterized by the work it does as "the voluptuousness of its sounds-signifiers."[81] In that sense, the grain goes beyond timbre, but also includes the significance created by the voice itself. Barthes's musings on the grain of the voice are—though explicitly grounded in the specific proximity of voice and language in the *mélodie*—not historical. His insistence on significance, and his understanding of the singing voice as "not personal" but nevertheless "individual," however, points clearly toward the ways in which the sounding materiality of the human voice is socially and culturally embedded. The modern voice, in all its grainy glory, whether plagued by pathology or honed to perfection, does have a history. It was invented as a tradition referring back hundreds of years, carried by petty anxieties over middle-class propriety, and mobilized to manufacture distinctions that were believed to inherently belong to the body, but needed repeated articulation to remain present.

CHAPTER 1

Seeing the Voice

The Emergence of a Field of Inquiry

In 1833, Marc Colombat de l'Isère published a rather curious book: a synoptic overview of statistics on "all sorts of stammering."[1] In it, the author presented research that must have seemed as peculiar to its contemporary readers as it does to us now. Who would set himself the task of mapping out stammering quite so extensively? But Colombat had an answer to that question too: he defined himself as an *orthophoniste*, a specialist in what is perhaps best defined as vocal health, as it was understood in the nineteenth century. The field of vocal health—whether it was associated with the neologism of ortophony or not—barely existed when Colombat coined the term, but would become a site of great intellectual, medical, and therapeutic activity throughout the 1800s. It embraced issues like stammering, lisping, aphasia, and other vocal practices we would now term speech impediments, but also questions of vocal beauty, of the health of the organs involved in speech (the larynx in particular), and of good vocal practice in a very general sense. This chapter attempts to chart how the diverse and ill-defined field of vocal health came into being and developed throughout the nineteenth century—without aiming to impose any clear analytical, medical, or aesthetic boundaries on it. Rather, I hope to elucidate how many practitioners—like singers and teachers—and scholars would participate (unwittingly, sometimes) in the establishment of an ambiguous field of research and practice that would subsequently become the breeding ground for processes of professionalization, medicalization, and specialization of vocal expertise in the latter half of the century. In doing so, I hope to lay bare what the voice was understood to be in the early nineteenth century, and how it appeared in various cultural and scholarly contexts. In their ambition to cultivate healthy, or even ideal, voices, various experts first had to establish what the human voice was, what it did, and what it meant—and this chapter follows the same line of inquiry.

The chapter also looks at the rising ambition of authors defining themselves as experts on the human voice to build a discipline that could be distinguished and recognized. It charts their attempts to fence off a field of expertise that included knowledge from different areas (medical as well as musical) but was nevertheless very exclusive toward what they considered pseudoscientists. And it attempts to understand their drive to ultimately see and professionally observe the voice. Central to these endeavors was a balancing act that we can already recognize on the title page of Colombat's statistical publication. It not only named ortophony as a potential field of inquiry, but also explicitly aimed to establish its author as a legitimate expert and part of an existing establishment of knowledge. The title page reads:

> Doctor, Founder and Director of the ortophonic institute for the treatment of stammerers and deaf-mutes; collaborator on numerous medical and scientific journals; member of the anatomical society of Paris, of the consulting committee of the society for intellectual emancipation, of the circle of surgeons of Montpellier, of the medical-surgical society of Lyon, correspondent to many philanthropic and literary societies, etc.[2]

Colombat's self-presentation as a self-made expert is complex, but fits into a wider practice of scholarship and medical expertise at the time. Many researchers, singers, speech therapists, and others sought this precarious balance between presenting their discipline as a new one providing innovative perspectives that would lead to new solutions on the one hand, and on the other, as one belonging to existing ways of knowing—and therefore deserving of the same respect as these older fields of knowledge. Three important practices contributing to that balance will be discussed below. First, the active and repeated distinction between legitimate work and what was seen as quackery or pseudoscience. Second, the use of vocabulary and literary references from the ancient world to lay claim to long-term solidity. And third, the invention of new techniques and machinery to exemplify innovation. None of these practices were straightforward. The line between real science in a new field and what was seen as less than legitimate was thin to begin with—and seems even more blurry with hindsight. Citing classic texts could show both ancient wisdom and what were considered the mistakes of the past now surpassed by modern research. At the same time, new inventions did not necessarily reflect the state of the art as authors perceived it—especially since there was no agreement on what could be counted among scientific invention and what could not. Overall, however, a picture emerges of the kind of experts that would cluster together and recognize one another's work—in tandem with the emergence of a visible picture of the voice.

Colombat's description of his own expertise is a particularly extensive one, but not atypical for its time or field, and did its share of cultural and representative work for its writer. The son of a rich merchant, Colombat certainly had the means to fund a medical education, but he had not been born into the nineteenth century's social networks of gentlemanly science. Rather than just adding a string of abbreviated degrees to his name—as some of his colleagues did[3]—he therefore chose to show his laboriously acquired belonging to the scientific fraternity more explicitly. His membership to various learned societies showed him to be a busy but above all a connected man who had a legitimate claim to medical authority as well as the gentlemanly sociability expected to accompany his professional skills.[4] The description showed how well-integrated Colombat had become, and also presented him as a daring innovator: the long list of societies and journals was mentioned also, perhaps, to compensate for the eccentric field the author had chosen for himself. As the "founder and director of the ortophonic institute," he posed as a highly specialized expert in a field of his own making, but with ties to the medical world in its larger sense. His ambitions seem to have largely been fulfilled: Colombat did become a widely cited, internationally acclaimed authority on stammering for much of the nineteenth century, even if his theories on ortophony are now largely forgotten.

As a highly trained expert in a developing field, Colombat had clearly felt the need to define and fence off his particular disciplinary claim to knowledge, but without an existing discipline to refer to, he invented one for himself and engaged in rather explicit practices of "boundary work."[5] With a specialization in issues that were commonly associated with quackery, he may have seen a particularly acute need to legitimize the authenticity of his expertise. Similar tendencies to overenthusiastically display qualifications were evident on the title pages of other works on stammering, which included, for example, James Hunt's self-identification as an "honorary secretary of the ethnological society of London. Foreign associate of the anthropological society of Paris, corresponding member of the upper Hesse society for natural and medical science, author of a 'manual of the philosophy of voice and speech,' etc, etc."[6] Presumably, this stress on formal education and inclusion in socially sanctioned networks of science served to distinguish "real" authorities from the shadier characters in the trade of stammering cures, who also populated the newfangled field of vocal health.

The elusive Mr. Fortuné Malebouche, for example, who became quite famous as someone carefully walking the line between public health expert and downright charlatan, did not present himself with any formal degrees (although he did stress the importance of his work being presented to the Académie des Sciences).[7]

All in all, Colombat's self-presentation on his 1833 title page prefigures and illustrates a process of developing a field of knowledge. It shows how a new field—of studies of the voice in general, as well as stammering in particular—consolidated itself by appealing to established sources of authority (such as formal degrees and existing societies), by distancing itself from questionable para-medical practices,[8] and by laying claim to innovation through the development of new methods of study and cure. The process often leaned on medical education and vocabulary, but was certainly not confined to the medical world. Degrees in music, new methods to study elocution, and attempts to unmask charlatans of all stripes all played their part in the establishment of an international, highly engaged network of experts concerned with the workings, health, and proper use of the human voice in the first half of the nineteenth century.[9]

In such a period of frantic energy, fierce competition, and high ambition, innovation and invention were on everybody's lips. Colombat was by no means the only one to come up with a whole new field—or at least with a neologism to designate the vaguely defined area of expertise that was starting to take shape around the human voice. And he was also not the only one whose grand neologism did not survive beyond its author's working life. In fact, only three recognizable innovations, and their associated terminology, gained any traction beyond the work of their original inventors: vocal hygiene, the speech center, and laryngoscopy. The development of the two former fields took place largely outside the field of vocal expertise, but they were picked up eagerly by the proponents of a physical study of the human voice. First, from the 1850s onward, but predominantly in the 1880s and 1890s, the hygiene movement gained popularity among vocal experts. Titles appeared claiming to study the "hygiene of song," "hygiene of the voice," "hygiene of the vocal organs" and the "hygiene of the orator."[10] Colombat, ever the innovator, was one of the first to pick up on the terminology in 1838,[11] but it quickly spread in France, Germany, and Great Britain, mainly among those with therapeutic ambitions. The second broad medical innovation to sweep the field was neurology. The "speech center" of the brain had already been identified by Broca in 1861, and a handful of experts discussed his discovery and its consequences for the study of speech and stammering in the second half of the nineteenth century. Nevertheless, it would take until the 1890s for authors identifying themselves as neurologists to join in the debates

Figure 2. William Gardiner, *The Music of Nature or an attempt to prove that what is passionate and pleasing in the art of singing, speaking, and performing upon musical instruments is derived from the sounds of the animated world. With curious and interesting illustrations*, London: Longman, Rees, Orme, Brown, Green and Longman, 1832, p. 76. Courtesy of the Wellcome Collection.

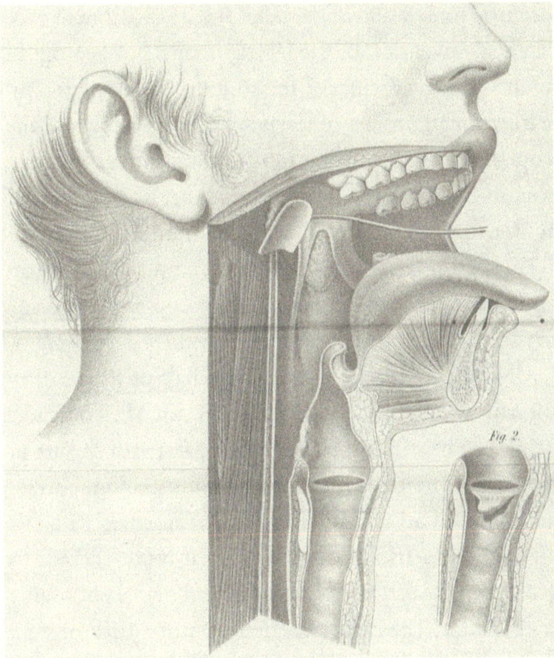

Figure 3. Viktor von Bruns, *Die erste Ausrottung eines Polypen in der Kehlkopfshöhle durch Zerschneiden ohne blutige Eröffnung der Luftwege nebst einer kurzen Anleitung zur Laryngoskopie*, Tübingen: Verlag der H. Laupp'schen Buchhandlung, 1862. Fig. 1. Courtesy of the Wellcome Collection.

about voice and speech.[12] Roughly around the same time, psychologists, who likewise took up the notion that speech and its pathologies resided in the brain, published their first theories on disfluency as well.[13] Speech pathology played an important role in the development and rise of neurology, but as the neurological approach toward speech (and above all, aphasia) led researchers away from the study of the physical production of voice in the throat, the two strands of research would eventually go separate ways.

Third, the rise of laryngoscopy (interior examination of the larynx) occurred in this time of higher specialization and strengthened specialists' claims to authority. They could now wield a very particular piece of equipment to signal their belonging to a select group of experts.[14] At the same time, however, the meteoric rise of the terminology of laryngoscopy and the appeal of its visual results also spread its basic tenets far beyond the scientific world. The second half of the nineteenth century also saw an expanding market for self-help guides and manuals for "Ärzte und Nicht-Ärzte" in which authors wrote "introductory remarks for the non-medical reader"[15] alongside their observations in laryngoscopy or neurology. Looking back at the field of vocal education and pathology in the nineteenth century, the rise of hygiene and neurology appear almost as temporary sidetracks. The central innovation of the field, and its subsequent evolution, was connected to Garcia's invention of the laryngoscope. No neologism rose in popularity so fast, and remained current for so long, as laryngoscopy. Manuals for the use of the instrument, results of laryngoscopic research and the clinical notes of laryngoscopists, started to appear in the 1860s and set the tone for increasing specialization, in both the medical and therapeutic branches of the study of the voice. The "renaissance men" who were involved in a variety of scientific societies gradually made way for medical specialists employed in hospitals such as the ear, nose, and throat clinic of Leipzig, laryngoscopic laboratories, and the phonetic laboratory in Paris, or as editors of the *Journal of Laryngology, Rhinology, and Otology*.[16] All-knowing teachers in the first half of the century had presented themselves in this rather broad fashion. A certain Mr. Fourcade, for example, termed himself a "Dramatic and lyrical artist. Master in the art of speaking, teacher of speech to deaf-mutes, professor of pronunciation, corrector of all vices of articulation and of all vocal defects."[17] But teachers who felt equally comfortable teaching song to artists and speech to deaf-mutes seemed to disappear from the debate on vocal health as educators in and owners of highly specialized establishments entered it. The early twentieth-century director who thought that his sanatorium for speech impediments was the "Einzige Anstalt dieser Art in Europa" simply did not acknowledge that at the turn of the twentieth century numerous institutions very similar to his own were established—as evidenced by the number of leaflets advertising them and manuals published by their founders and employees.[18] Others went in the other direction, taking a democratic approach, in which they addressed their findings to "all who read and speak in public," "orators, clergymen, vocalists, and others," "speakers, singers and all who love the arts of speech and song" or "singers, elocutionists, educators, barristers, clergymen, and to all others desirous of having a pleasant voice and good health."[19] Manuals, specifically, were written for an increasingly diverse audience, and experts on disfluency and song even explicitly framed their work as useful for *Selbstunterricht*, for "dilettantes" or "those who are unable to submit themselves to treatment by a speech physician or an Institution for stutterers."[20]

Stout of Teeth, and Strong of Tongue

The field of speech pathology—and of stammering especially—was particularly vulnerable to allegations of fraud and quackery. Or, as many practitioners understood the situation in the first half of the nineteenth century, their field was particularly porous and too open to, or indeed overrun by, unconscientious merchants of miracle cures. The plethora of unreliable stammering cures was, according to some authors at least, the direct result of the lack of serious research and thus of trustworthy knowledge about the subject. According to Richard Cull, who called himself a "tutor in elocution," "The quackery and mysticism in the cure of impediments of speech arises from the general ignorance of the philosophy of the human voice and speech. Explanations of their phenomena have been demanded, to administer to the necessities of defective utterance, and to satisfy the ear of taste in a just elocution, but in vain. Bread has been asked for, and stone has been given."[21] His short refutation of all this quackery and mysticism, published as a "letter to George Birkbeck," received a rather favorable review in the *Medico-Chirurgical Review* of 1836.[22] The reviewer noted that Cull's polemic tone "had given his style of writing somewhat of King Cambyses' vein, or a little of the swagger of Ancient Pistol," characterized the text as a "pamphlet" even though he largely agreed with its sentiments, and applauded Cull's efforts to delegitimize any simple cures promising swift results. He also agreed with Cull's cultural diagnosis of the fashion for quackery and its roots in a lack of knowledge. The review starts with an evocative reiteration of this problem:

> It would be tedious to enumerate the various modes that ignorance, empiricism and imposture have at various times proposed for the cure of impediments of speech. From the farmer in Joe Miller, who made his son sing—"Daddy your house is on fire," down to Mrs. Leigh, the Lady in New York, who forces her disciples to keep the tongue against the teeth uninterruptedly for three days, the plans have been innumerable, the proposers confident, and the success, of course, immense. Yet, singularly as it would seem, although each plan is perfectly successful, the one that follows is advanced expressly, because no such thing as success can be obtained. How this paradoxical state of things may be explained, we leave to others to determine.

Conflating fraud ("imposture"), unconventional methods of study ("empiricism," which was also known as "sensualism"), and plain old ignorance, the anonymous reviewer does not use the term quackery, but refers very effectively to its assumed preponderance among experts on speech impediments.[23] Moreover, by evoking a linear chronology of one infallible but ultimately useless cure following another, he also underlines the long history of the connection between quackery and speech impediments. Evidently, the story of the self-declared miracle workers of stammering was understood to be familiar to readers of the journal. The legitimacy of any work referencing the cure of stammering was therefore so inherently fragile that it needed to be preemptively defended. As Cull's reviewer noted: "his [Cull's] practice [could be] tainted with the suspicion of empiricism"—in other words, it seemed to be based almost exclusively on sensory experiences, rather than rational thought and research.[24]

Experts across Europe felt the same need to distance themselves from the fraudulent practices that were often associated with the study and cure of stammering. German physician F. Angermann, writing for "physicians and pedagogues," stressed in the preface to his manual that his was a "scientific" study of something that "until then had generally been surrendered to charlatanism of the highest order."[25] Likewise, father and daughter Vernet, who are both designated as authors of "theoretical and practical lessons" on the subject of stammering, deplored the fact that "charlatanism has, until now, defined the treatment of this infirmity."[26] By doing so, these experts acknowledged the fragile status of their new field, but also claimed a degree of distinctive trustworthiness for their own innovations.[27] Central to this need to distance one's work from the widespread practices of charlatanism was not only professional pride in the work itself, but also an almost emotional attachment to the identity one could construct as a particular kind of expert. Joseph Poett, in a particularly explicit phrase denouncing any kind of alternative knowledge, hinted at the visceral quality of the expert's identity: "The above particulars may seem strange to enter into," he noted, "but I think it necessary to insert them, in order to prove that not the slightest degree of quackery, or particle of charlatanism is embodied in any part of my treatment of the disease."[28] His need to prove his credentials by providing particulars can perhaps be read as a precursor to what would become typical of the scientific method and the necessity of experimental duplication. More interestingly, however, the phrase also points to notions of disinterest and distancing (no charlatanism is embodied, i.e., physically rooted, in his methods), while simultaneously containing a great deal of keenly felt anxiety.[29] He was hardly the only one to get emotional over the subject of legitimacy and quackery. In Hartley's *Observations on the Different Kinds of Impediments of Speech,* the emotional attachment to a construction of self as a trustworthy and legitimate authority took the form of righteous anger, directed at both the quacks themselves and their patients/clients. "The difference between those who profess such barefaced quackery, and those who believe them, is, that the former are knaves, and the latter fools; and deserve to be fleeced, if I may be allowed the expression, of the exorbitant sums which they pay for their credulity."[30] Despite Hartley's strongly worded opinions, however, quackery was not clearly defined or visible in the field of vocal pathology in the first half of the nineteenth century. (And that is perhaps where most of the strong need to distance oneself from fraudulence derived from.) Between the 1830s and the 1860s, different methods to gather, analyze, systematize, and eventually apply knowledge about speech impediments in particular and the voice in general would be developed and compete for legitimacy.[31] As researchers on counterhegemonic medicine have suggested, it is helpful to think of these conflicting theories not as a linear development toward a more scientific (or disenchanted) understanding of the world, but rather as players in a "medical market," with elements of economic and cultural competition.[32] For the field of voice pathology, which had not quite found its place within conventional medicine but also cannot be entirely relegated to the counterhegemonic sphere in this period, the notion of a double market works remarkably well. On the one hand, as Hartley's sneering description of dealings between fools and knaves suggests, the sales of stammering cures could be a lucrative business, and experts, regardless of their methods, competed for the attention and patronage especially of middle-class stammerers. On the other hand, as the emotional declarations of authors like Poett, Angermann, and Vernet show, they were also engaged in a market-like "struggle for recognition," seeking not only the "trust, belief and esteem" of their patients, but also that of their peers.[33] That entailed an ongoing process of practice, lectures, and publications.[34]

The struggle for recognition was a reasonably public one. Although the actual care for patients and their eventual cure could be an isolated, almost invisible practice, the construction of one's reputation as a trustworthy expert and an effective healer depended on how one was perceived at scientific society meetings, in publications, and even in the press. Many of the brochures and monographs I use in this book should probably be read at least in part as the public articulation of their authors' expert personae,[35] as a way to navigate a changing field of practice and knowledge in which reputation mattered and the expectations toward a speech-practitioner changed considerably over a short period of time. One particular practitioner, active as a purveyor of a cure for stammering in (mainly) the Low Countries and France in the first half of the nineteenth century, can serve as an example of this gradual, and at times chaotic, process of professionalization and legitimation: Fortuné Malebouche. In 1827 his method to cure stammerers was presented before the Académie des Sciences by François Magendie and Constant André Duméril. In a booklet published in 1828, Malebouche articulated this moment as a turning point for his method and himself: in *Nouvelle méthode pour guérir le bégaiement*, he described the difficulty of establishing himself as a practitioner without any formal credentials ("aucun corps savant ne s'était prononcé en France pour la nouvelle méthode").[36] Moreover, he noted, presenting his method before the academy without the support of any institutions or journals was risky and "quite unpleasant." What would happen if the report failed to convince the academy? Luckily, despite all the obstacles that might confront the independent practitioner, the various types of proof and the handful of recommendations that he could supply to the academy convinced its honorable members, and "the beauty of the method has overcome all obstacles." Malebouche's institution to cure stammerers in Paris was set to thrive, and in 1841, he penned another short book on the subject, entitled *Précis sur les causes du bégaiement et sur les moyens de les guérir*.

At first sight, Malebouche seems an unlikely figure to exemplify the search for legitimacy and the creation of a respectable field of knowledge. His background as a practitioner is decidedly unclear, as are his family connections and even his place of origin. He just suddenly appears on the scene of stammering cures, under what is most likely an assumed name. Fortuné Malebouche, it seems, named himself after a figure in one of the most famous pieces of medieval French literature. In the *Roman de la rose*, Malebouche is a slanderer, a *losengier* whose language represents "the opposite of courtly discourse,"[37] and whose speech consists mainly of the abuse of language. It is an interesting choice for a pseudonym: on the one hand, the simple term Malebouche brings the physicality of speech and its location in the mouth to mind—pointing to the tangible ways in which stammerers' speech is unlike proper speech. It represents faults in the manipulation of words in the mouth. On the other hand, the character of Malebouche in the *Roman de la rose* is somewhat unsavory and above all unconventional. In opposing the norms and expectations of courtly speech, Malebouche can, in some ways, speak truth to power.[38] In the 1828 brochure on the presentation of his work at the academy, that is the role that the modern Malebouche seems to claim. Lamenting his lack of support from existing institutions also allows him to distance himself from medical science's established order and leaves him free to criticize both (which he does more fully in his later writings in 1841). In a sense, the Malebouche of 1828 sets himself up as a mysterious, quack-like figure, a hero in the romantic story about the origin of his method. His brother (yet another unidentifiable Malebouche) had somehow acquired the method from a certain Mrs. Leigh.

> Madame Leigh, a New York resident widowed at the age of 36, was welcomed into the family of Dr. Yates with kindness and received the most selfless care. One of the daughters of the doctor, aged about 18, was

> suffering from a fairly heavy stammer. Mrs. Leigh thought she could not better express her gratitude to her hosts than by relieving this lady of her infirmity; to this end she read all English work related to stuttering. Not obtaining the required result from this study, however, she put herself to observing with much perseverance the nature of the infirmity she wished to cure herself, on the affected subject.[39]

As studying existing methods did not seem to work, the widow came up with her own. After having treated the good doctor's daughter, she established a practice for stammerers in New York, applying her new method with great success. The story becomes even vaguer after that point: "The inventress of this method, encouraged by her American success, wanted to spread it in Europe and entrusted it to M. Malebouche, the brother of the one now in Paris."[40] With the assumed name, the unclear (and exotic transatlantic) origin of his cure, and the hyperbolic stories of swift and complete cures, Malebouche bears all the hallmarks of a classic quack. He may not have been selling potions, but he was touting something of a miracle cure. Moreover, stammering cures were a known trope of quackery.[41] What made his method most suspicious, perhaps, was the insistence on secrecy. As Magendie and Duméril made clear in their reports, they had witnessed only the results of Malebouche's cure, not its inner workings. At the end of their report, they noted their regret at the refusal to divulge the methods of the cure they had observed: "Your commissioners declare that, by the method of the widow Leigh, a cure for stuttering can be achieved ... but they cannot help expressing their regret that the inventor of this method has not thought it necessary to comply with the established usage nowadays, to make public all discoveries that can become profitable to mankind."[42] In his 1828 introduction to the published report, Malebouche defends this choice for secrecy. The method is, he notes, not easy to transmit and cannot be divulged with sufficient clarity to be useful.[43] The 1841 text, however, displays a very different discourse on secrecy. According to this version, it is the mysterious Mrs. Leigh who had insisted so strongly upon keeping the method hidden. Malebouche himself had, he wrote, always thought that more transparency would be the better and more scientifically proper strategy.[44] In this account, the report before the academy is a first step toward openness, after which publications in the *Dictionnaire de Médecine* (by Magendie and himself) gave the method more notoriety and more legitimacy in conventional medical circles. Although the initial observations at the base of the cure are still attributed to Mrs. Leigh, the techniques leading to an actual cure are no longer presented as a method invented and practiced by her, but rather as "the method we have made public."[45] As the method moves deeper into a territory of medical respectability, Malebouche claims more authorship for its methods and even the observations in which they are rooted.

And yet, in his new role as a transparent author, keen to thoroughly set out his work, analyze that of others, and take part in scientific debate, Malebouche still retains some of his initial more rebellious and unconventional characteristics. His new pamphlet is, once again, written (ostensibly at least) from the margins, arguing against what he presents as the consensus of the medical establishment. In *Précis sur les causes du bégaiement*, he states the wish to "explain our position and that of other practitioners clearly."[46] What follows is an extensive, if not necessarily precise, critique of existing theories and practices surrounding the cure of stammering. Central to his discourse are the operations on the tongue carried out at the time by, for example, Johann Friedrich Dieffenbach.[47] In Malebouche's account, these surgeons are described as the rule, representing existing medical practice as a whole.[48] The author himself comes out of the narrative as the bringer of a completely different method, the clearheaded scientist who has thought through his observations more fully and therefore is capable of going beyond what he calls "this mania of engaging in practice without having seriously studied the theory."[49] Almost half of the *Précis* consists of scathing criticism of these operations, which according to Malebouche had no tangible results. His own method is then presented in contrast to this ostensible consensus on surgery, not only as the cure that, unlike surgery, actually works, but also as the method that draws very different conclusions from the basic observation attributed to Mrs. Leigh, that is, the faulty position of stammerers' tongues when they are not speaking.

In fact, surgery on the tongue to cure stammering was not the established consensus at all—surgeons like Dieffenbach and Amussat and their practices were quite controversial. Moreover, Malebouche's own method of somehow pushing the tongue upward in the mouth was not unique to him either. Yet, although Malebouche was not quite the unique innovator he perhaps wished to be, he also was not the charlatan he appears to be at first sight. What looks like peddling a miracle cure did to a degree grow into an attempt to contribute to public health in a fashion similar to the sanatoria for lung diseases that were springing up around the same time, and his concerns surrounding speech impediments were shared by many physicians and politicians of the time.[50] Moreover, in presenting his method, Malebouche draws upon a number of techniques and narratives that, throughout the century, would become typical of the increasingly proper scientific discourse displayed in the field. His work and backstory were more colorful than that of most of his colleagues, but in many ways his story is quite representative of the rise of a therapeutic method for speech impediments—or indeed a move from outright quackery to a reasonably respectable and systematic practice surrounding vocal health and beauty. In the 1841 text, he explicitly represents himself as a modern scientist, looking for truth and proof—unlike practitioners of the past who lacked this drive. In his *Précis*, he speaks of a time when "there was an enormous tendency not to take previous experiences and proven precedents into account, and a no less marked disposition to despise theory."[51] The account he gives of his own method is a very systematic one, an account of a theoretical study regularly interspersed with moments of empirical proof and demonstration. He describes a number of solved cases to support his argument, but also calls upon the reader to repeat his empirical findings, which "one could," he noted, "easily verify on oneself."[52] Most typical of the developing scientific style of writing about stammering, perhaps, is his mentioning of Demosthenes. Malebouche's account very clearly places the ancient orator in the less systematic and unscientific past, but in bringing him up, Malebouche does manage to hint that his own knowledge of the classics was more thorough than that of his competitors who believed that stammering could be cured by placing stones in one's mouth.[53]

With this range of different scientific, intellectual, and erudite narratives, Malebouche's 1841 brochure prefigures the work that would follow in his field of interest, and the two-pronged path practitioners in this discipline would follow. On the one hand, insistence on empirical, experimental, and comparative research led to a large range of new techniques and methods that were explicitly represented as innovative and modern (such as the use of statistics, or various new ways to visualize sound). Authentic observation was central to this impetus toward a new scientific way of thinking about the voice, and was represented as a long overdue innovation. According to Richard Cull, for example, the public had long clamored for a serious study of the reasons and grounds for stammering. Exactly who had been calling for this research is not quite clear—Cull's use of the passive voice cleverly obscures any sense of agency and authorship in the field of vocal health preceding his own work. He notes that "explanations . . . have been demanded," not who demanded them.[54] Dismissing the recent past, Cull rather obliquely refers to the knowledge of classical authors in much the same way as Malebouche. He suggests the depth of his own erudition by calling upon a vague, ancient precursor to his field of study, but makes the superiority of new scientific methods over the ancients' unprecise knowledge very clear: "The subject called for examination by accurate observation and experiment, not for the transcription of Greek and Roman knowledge, as if it were impossible for us to do more than quote classic authority."[55]

Actio, Actio, Actio

References to classical wisdom—and how it could be adapted and surpassed in modern times—reappeared in scientific treatises, educational manuals, and therapeutic reflections throughout the nineteenth century. The knowledge of the ancient Greeks, in particular, was thought to be relevant to studies of the human voice because of the high status and importance of public speech in the Greek polis. In the first half of the century, specifically, as fields like laryngology and vocal physiology were still taking shape, authors presented their work partly as reviving or recovering the lost art of the ancients. Some referred back to the orators of yore as the last torchbearers of a great culture of the spoken word that had been lost almost completely and needed to be painstakingly reconstructed. This was particularly the case for authors who focused on eloquence, but the loss of the spoken word was connected to the faulty elocution of modern times as well. According to French rhetorician Paignon, the fall of Rome was directly connected to the loss of great public speech: "The moment Demosthenes' great voice was heard, Athens was saved. Cicero spoke, Catilina went. Rome perished when eloquence degenerated."[56] Physicians and physiologists were slightly more critical of their classical predecessors (pointing, e.g., to their limited knowledge of anatomy) and presented Greek and Roman texts not so much as wisdom to be reconstructed, but rather as the cradle of a long-standing, if somewhat meager, tradition of studying and training the voice. For James Mackness, for instance, the great and impactful speeches "in the ancient republics of Greece" were the first instance of a long series of examples of effective public speech. "To the eloquence of the Forum and the senate succeeded the eloquence of the pulpit, and Paul, Chrysostom, Cyprïan, Augustin, and a host of Christian fathers and preachers."[57] Great speakers such as Demosthenes and Cicero were therefore the precursors to those that the readers of Mackness's *Dysphonia Clericorum* would be more likely to aim to imitate. The manual was expressly designed for clergymen who had (physical) trouble speaking, and it comes as no surprise that great, religiously inspired speakers were quoted next to the ancient and modern political ones. "The voice of Luther" appeared as a particularly powerful one; it was "the potent spell which roused the enchanted sleepers in fabled story."[58]

Those not addressing clergymen with sore throats chose generally more classical, and very well-known, examples of wisdom of the past to cement their claims on the importance of the human voice for health and political culture, appealing expressly to their audience's classical education. Four great men of ancient Greece and Rome were particularly popular: Aristotle, Demosthenes, Cicero, and Quintilian. The choice of examples overlapped, at least partly, with the curriculum most schoolboys would have been familiar with (a curriculum that, despite the lamentations over lost art, was very much geared toward developing skills and styles for public speech).[59] Moreover, Aristotle and Quintilian had both famously written, in a sense, scientifically about the voice. They focused on physiology and eloquence, respectively, and therefore provided modern scientists not only with deep roots to legitimize their research, but also with a foil against which they could demarcate their own more experimental endeavors. The "imperfect analyses and vague treatises that the ancients have left us on the subject of the voice"[60] made for a perfect starting point for physicians, therapists, and elocutionists alike. Demosthenes provided the best stories to start off an investigation, a diatribe against quackery and a vigorous call for self-discipline and self-cure, and was presented to readers in all genres of works concerning the voice. Most middle-class men would at least have heard of him: Plutarch's "lives" included fascinating stories of the orator's rise to power and influence, and in the age of the self-made man the story of the youth with a speech impediment transforming himself into a powerful speaker through mere training and willpower was bound to be a popular one.[61] The "pebbles of Demosthenes" were so well known that they could be referred to without any further explanation—they were cited almost as a saying. "To be an orator," Charles Rivail noted at the end of the century, "one always has to, more or less, have chewed on Demosthenes' pebbles."[62]

Although the life story of Demosthenes, and especially Plutarch's retelling of it, was obviously well-known to readers across Europe, authors retold and refashioned the story in remarkably different ways, molding it for their own purposes. For authors of elocution manuals the Athenian statesman was "the prince of orators," one of the most eloquent among his peers and therefore rhetorically—and thus politically—successful.[63] His prowess as an orator, often mentioned in one breath with that of Cicero, was uncontested. His methods for scaling the heights of Greek eloquence and public speech were the subject of debate, however, as many authors' own interest in the speaking voice rather than rhetorical style led them to an interest in vocal production over discourse.

As they kept repeating with Demosthenes, speech was all about actio. And it is in actio (delivery) that Demosthenes was such a fascinating and problematic figure himself. Louis Dubroca was one of the authors reminding his readers of Demosthenes's answer to the question of what the three most important pillars of eloquence were. It's "Actio, actio and actio," a response that according to Dubroca showed that the orator who overcame a speech impediment gave ample attention to pronunciation.[64] The can-do attitude displayed in the story that—via Plutarch—was passed along in various scientific and educational manuals retained its appeal throughout the nineteenth century. Laryngologist Henry McCormac addressed its enduring popularity in his scholarly work on stammering: "The Demosthenic method seems, of all others, to have been the greatest favourite with mankind, since the time of the celebrated orator, from whom it has its name; so fascinating is the sanction of great authorities, that this method has been advised by writers and practitioners in almost every form of impediment of speech, whether resembling or not, any one of the different ways in which Demosthenes is supposed, by different authors, to have been affected."[65]

Here was a sufferer of a speech impediment who had effectively cured himself—thus proving what most stammering experts of the nineteenth century professed. A cure was not only possible, but also attainable. In casually quoting Demosthenes's habit of "stop[ping] daily by the coast where, his mouth filled with pebbles, he shouted at the waves," some authors seemed to suggest that simply reproducing that practice would relieve modern speakers of their stammer as well.[66]

Other specialists—most notably those who specifically addressed stammerers—were more hesitant. Simply "chewing pebbles" or shouting on the beach sounded, of course, suspiciously like a miracle cure and did not fit the moral and therapeutic norms of the time. Yet for them, as well, Demosthenes provided an excellent way to show the possibility and validity of a cure dependent not on the mechanical/material means the orator used, but on his discipline and perseverance. Although the ancient Greeks had made many errors in their efforts to understand the voice and its pathologies, the story of Demosthenes could still be an inspiring and hopeful one. "The oft-cited example of Demosthenes should strengthen their [stammerers'] courage and encourage them to use all means to give the movements of their organs the strength and precision they lack. Plutarch recounts with detail and in the most fascinating manner how the great orator succeeded to rid himself of a stammer that seemed destined to exclude him from the rostra forever."[67] In fact, based on critical reflections on Greek research and knowledge, modern stammerers could be even more optimistic. The new methods made available to them by the superior study of modern experts made a cure *more* likely than had been the case for the great orator. All they had to do was to match his "firm and perseverant will power."[68]

Throughout the century, elocutionists and other voice therapists would try to strike a balance between holding up Demosthenes as a shining example for all those desperate for a cure, and criticizing the "legend" that such simple means as pebbles (which bore great resemblance to the tools peddled by early nineteenth-century elocutionists, such as Itard) could lead to long-term recovery. A number of authors even questioned if Demosthenes had a stutter in the first place, noting that a careful reading of "the ancients" (presumably Plutarch) revealed less severe impediments. The overall consensus was, however, that Demosthenes had somehow stumbled upon a cure for stammering that could only later be properly understood after careful study and consideration by modern observers (who had the benefit of greater anatomical and physiological knowledge). Claude Vernet summed up the somewhat ambiguous place of the story of Demosthenes in nineteenth-century vocal science. A question attributed to an innocent girl shows the fame of the story, its aspirational strength, and the suspicion it raised in modern scientific minds.

> *CORALIE*. Daddy, is it true that Demosthenes managed to cure himself with such simple means? It is not very probable!
>
> [*Answer*] My child, everybody has believed in this type of cure, and we would still believe it if the natural method would not have been inherent in it; but whoever knows the organic cause and mechanism of stammering will also easily recognize that the little pebbles Demosthenes used were but secondary or ephemeral means in his treatment, and that his habit of speaking alone in solitude and exercising vigorously and without pause, together with a severe routine that Demosthenes was forced to observe living in the woods and on the sharp rocks were the most important elements of this beautiful cure.[69]

Demosthenes's qualities as an eloquent, stylistically brilliant orator mainly served to underscore the great utility of strength and endurance, and therefore showcased the possibilities that "cures" for speech impediments could entail. However, other figures of classical eloquence invoked by nineteenth-century authors were more explicitly cited as masters of eloquence. Their works—and Quintilian's *Institutes of Oratory* especially—were mobilized by modern experts to further an educational approach to both rhetoric and vocal production. The great Roman rhetorician's introduction to his *Institutes of Oratory* held great appeal for vocal experts at a time when the voice was increasingly interpreted as a material, corporeal tool for the practice of speech. Quintilian's reflections on oratory were remarkably "holistic," containing notes on pronunciation as well as style. Most importantly, he saw a higher education in rhetoric in a larger context of schooling, starting at a very young age. In his first chapter, Quintilian exhorted fathers to have great ambitions for their boys: "The man who shares this conviction, must, as soon as he becomes a father, devote the utmost care to fostering the promise shown by the son whom he destines to become an orator."[70]

This notion, that to teach men to speak well one had to develop every aspect of childhood education, was remarkably well-suited for adaptation to nineteenth-century mores and values. Quintilian's attention to the child's first sounds, for example, became a call to include elocutionary training in women's colleges because it would prepare them for their important task (as future mothers) to mold the vocal organs of their children. "It is perhaps even more urgent to introduce the foundations of good pronunciation and good reading at schools for young ladies," as Duquesnois, the author of a manual for orators and readers, phrased it. "In fact, do not young ladies have a calling to become mothers, and in that capacity to raise children? 'It is on the care they take in forming our first sounds that we depend for our good or bad pronunciation', Quintilian says."[71]

If Demosthenes was held up as an example to patients, or readers of self-help manuals, Quintilian served as a glorious predecessor to the authors of studies on both eloquence and vocal hygiene. His *Institutes of Oratory* provided a classical example of the kind of holistic and educational approach to voice most authors aspired to, and his work as an orator and educator seems to have prefigured and legitimized the professionalization of these practices in the nineteenth century. Hygienist Auguste Debay presents him as an example of an expert on the study and development of the voice in ancient Rome, but also as proof that "in Rome, vocal art had become a profession as lucrative as it was respectable."[72]

Debay's work on the hygiene of the voice and "gymnastics" for the vocal organs thus harks back to ancient times (he manages to mention not only Quintilian, but Demosthenes, Cicero, and Aristotle as well) in order to root a study that is very much steeped in the newest buzzwords of the day. At a time when elocutionists and hygienists were struggling for scientific professional recognition, when institutes for vocal training were slowly freeing themselves from the association with quackery, and when a corporeal approach to rhetoric entered the medico-scientific and educational mainstream, setting up a financially viable business as a voice trainer was likely to be treated with suspicion. Quintilian's work showed that such an endeavor could be not only "honorable" but also effective and even necessary. According to James Mackness, the Roman's work also proved the need for vocal cultivation. "It is true that no cultivation will make a bad voice a good one, but the words of Quintilian—*Augentur autem sicat omnia ita vocis quoque bona cura et negligentia minuuntur*—are as true now as they were on the day they were written."[73]

The "princely" orators and great minds of bygone times thus served to legitimize what was otherwise presented as a thoroughly modern approach to vocal health and excellence: an educational or therapeutic practice based on early intervention and self-imposed discipline. Their knowledge and recommendations were not accepted wholesale, however, and the gaps in the Greeks' anatomical knowledge were a returning object of critique. Aristotle, whose work on animal voices was cited regularly, was mobilized to insist on the exclusively "human" character of speech, yet was roundly refuted in most areas of anatomy and physiology. Both his *Politica* and *Historia Animalium* were explicitly mined for phrases that suited the sociocultural rather than the medico-scientific agenda of nineteenth-century social research. And it is perhaps because of Aristotle's insistence on the specifically human character of speech, and because of his engagement with categories such as gender and age, that he was read so eagerly—and criticized more thoroughly than other classical authors. Even if many authors included the Aristotelian dictum that man is a political animal and distinguishes himself through the capacity for speech, his general understanding of the voice seems to have been labeled, as Édouard Fournié framed it, "wrong, but eminently logical."[74]

Like other scientists across Europe, Fournié focused on Aristotle's distinctions within the animal kingdom pertaining to voice—a notion that was generally recognized as the starting point for both physiological and acoustic theories of the human voice, but also one that led to a host of false assumptions. (Most notably, Aristotle's observation that "those animals only which possess lungs have a true voice, it is, therefore, confined to the first three classes of the animal kingdom"[75] was partly disproved. The long-standing Aristotelian misunderstanding that voice was produced by the movement of air was rectified in numerous works on acoustics.[76]) Far more important to nineteenth-century scientists, however, was Aristotle's interest in physiological distinctions and their audible articulation in mammalian voices. The observation that, unlike almost all other species, bovine animals gained higher voices as they aged and that cows had a lower bellow than bulls drew particular attention. The systematic study of gendered and age-related differences in (human) voices became highly popular in the nineteenth century, and Aristotle's early work on animals fit with the "modern" push to understand and explain these differences in natural/biological terms. Even his explanations for these vocal divisions of age and gender could be recuperated for the dichotomous understanding of male and female, adult and young, sexually active and "eunochoid" voices of the nineteenth century. Even if "the formation of tones, modifications of the voice according to sex, age, health or illness" were not discussed in *De partibus animalium*, Fournié noted that they had been addressed in other work.[77]

Fournié's engagement with Aristotelian notions of humanity, his adaptation of the ancient philosopher's conclusions to modern conventions of identity, and his reliance on classical sources to bolster his own image as a professional are a reflection of a more general tendency among his colleagues at the time. His 1866 *Physiologie de la voix et de la parole*, however, contains a much more thorough and systematic treatment of his predecessors' work than was usual at the time—and in a sense prefigures norms of professional/scientific practice of the later years of the nineteenth century. Rather than drawing upon a number of literary quotes and anecdotes to suggest the author's culture and erudition, Fournié's *Physiologie de la voix et de la parole* includes a well-structured account of the state of the art of what we can therefore call his field or discipline. Entitled "History and Critique of the Different Theories of the Voice," the chapter does exactly what we have come to expect in scholarly work. It offers a well-thought-out overview of existing research, presented and evaluated by the author, who frames his own work in the context of what other "experts" have claimed, while weighing different methods, theories, and results against each other.[78] Fournié noted that his thorough examination of existing work was unusual. "The uncommon space we have granted to the historical and critical part of our work deserves to be justified," he noted at the start of his chapter.[79] His justification was mainly rooted in what he claimed was the very chaotic and confused state of knowledge on the voice: old and new theories circulated freely, contradicting each other without sufficient examination and debate. Even though the available material was overwhelming (as Fournié noted, "the task of the critical historian is not always easy"[80]), it was necessary to delve back into history and wade through all these perspectives before embarking on a study of one's own. Fournié did indeed go back very far, and had obviously endeavored to provide an exhaustive rather than a selective overview, noting that: "One cannot, indeed, gather too many historical documents on all issues at a time when, in response to a general need, we are busy building a medical history on a more intelligent, more rational base."[81] The chronology of all these documents starts, for Fournié, with Hippocrates and then dwells on the knowledge of the ancient Greeks for several pages. Hippocrates, Aristotle, and Galen all appear as thinkers who essentially first raised the questions Fournié still considered important in his own research. Hippocrates prefigured medical research on the voice, Aristotle drew attention to the voice's capacity to distinguish people from animals and each other,[82] and Galen was the first to compare the human voice to instruments (an issue that would fuel discussions throughout the nineteenth century).[83] After this rather thorough engagement with the classic authors, Fournié more or less dismisses the Middle Ages, claiming that "after Galen, one has to go to Fabrice d'Aquapendente [1537] to find an author who has occupied himself seriously with the vocal organs."[84] His sixteenth- and seventeenth-century examples of vocal experts are generally represented as erudite scholars with great knowledge of Greek and Roman science and culture. He refers to Mersenne, for example, who according to Fournié "was well-versed in the work of the ancients."[85] By far most of Fournié's attention goes to what he calls "sages of our age."[86] More than half of his history consists of a thorough critique of his contemporaries; and apart from one German author, all the contemporary sages that he discusses were French physicians and anatomists. Although the stated goal of the history was to dispel confusion and create clarity in a long-standing field of study, the final conclusion of the chapter reveals a more pragmatic purpose to this scholarly labor: the chapter showcased the breadth of his knowledge, its roots in the past, but above all its author's capacity for innovative and experimental work. It announced the kind of scientific practice that vocal experts would increasingly undertake in the second half of the nineteenth century: "Our investigations have extended, as has been seen so far, both to the physiological and the anatomical perspective; we are sufficiently familiar with the handling of the laryngoscope; we have looked, in a word, to gather all favorable conditions for the production of a useful and conscientious work."[87]

Making the Invisible Visible

Fournié gave ample proof of his scientific credentials and his acquaintance with the laryngoscope throughout his work, dedicating a whole chapter to it in his work on vocal physiology. As announced in its title, *Physiologie de la voix et de la parole* had "figures dans le texte," being richly illustrated with different representations of the vocal organs. For all his interest in classical knowledge and his concern for a legitimate connection to research that had been done in the past,

Fournié was an observant scientist first and foremost. His observations—despite their focus on an acoustic phenomenon—were above all visual and could therefore be best represented by the extensive use of images. *Physiologie de la voix et de la parole* contains representations of male and female, young and old, healthy and ill larynxes, a number of anatomical drawings to represent the mechanics of voice production to the reader, and representations of the laryngoscope and its use.

The voice that Fournié represents is an eminently material one: rather than trying to represent the invisible aspects of the voice (i.e., its sound), he is laying bare its hidden but nevertheless very tangible features: the workings of the vocal folds, the shape of the pharynx, the vibration of the vocal cords, and so on. Reproducing the intangible features of the voice was of course largely impossible at the time: inventions such as the phonograph were being developed around this period but were unknown or unavailable to the public and scientists alike. Nevertheless, Fournié's focus on revealing the hidden rather than the invisible aspects of the voice seems to be a choice more than a necessity (rather tellingly, he does depict an artificial larynx for his readers). His choice appears to be the result of the radical shift toward the tangible in studying the human voice. For centuries, the hidden and eerie characteristics of the voice had lent it an air of mystery, but scientists—like Fournié—who were active after the first experiments on dead human larynxes were committed to lifting this veil. The first steps toward this materialization of the voice had arguably been taken in the eighteenth century, by people like Wolfgang von Kempelen, who not only observed the voice but set out to build an artificial one, mimicking the mechanism of the vocal organs. He described his research and inventions in his 1791 publication *Mechanismus der menschlichen Sprache: nebst der Beschreibung seiner sprechenden Maschine,* in which he expressed his certainty "dass die Sprache nachahmlich sein Muss."[88] Despite such optimism, however, experiments on mechanical, artificial, and dead larynxes led to unsatisfactory results. They sounded unpleasant and awkward. Clearly, there was something about the living voice these experiments had failed to demonstrate.

Although the produced sound was used as a way to establish experimental success, and to judge the vocal characteristics of real live larynxes as well, it was not scientists' first concern: the sound of the voice remained largely the domain of singers, actors, and other (professional) speakers. In order to represent the audible aspects of the voice, these experts generally relied, at least during the first half of the nineteenth century, on methods that had been central to their trade for centuries. Musical notation, for example, was used to great effect to represent those sounds in which pitch and rhythm were considered essential.

Musical notes could represent the "puling sound of a spoiled child," for example, or different kinds of birdsong, or even "a young lady calling her lover."[89]

Using notation to lay bare the melody of children's, women's, and animals' voices drew attention to these voices' esthetics and presumed natural characteristics. Written language was used to support these notations, adding enunciation to the practices documented, but the stress remained firmly on sound more than content (or on phonos rather than logos, as Barthes would have it).[90] For the speaking voice—which was imagined as almost exclusively male, as it was so closely tied to professions in which public speech was central—simple transcription through written language sufficed for most speakers and readers for the first half of the nineteenth century.[91] And even later on, as acoustic recording was developed and became more available, the ability to render the sound of the spoken word on paper was a matter of a writer's skill: readers of newspapers, columns, and novels were used to reading elaborate and imaginative descriptions of the vocal aspects of real and imagined speech.[92]

Both written language and musical notation were of course old methods to reproduce vocal sounds and utterances on paper, and although both notes and the alphabet represent sound, they do so in very indirect ways. The letter "a" does not preserve a particular sound—it merely encourages the reader to produce or recall a number of sounds that can be represented by "a" (and depending on the language of both writer and speaker, their recollections might differ).[93] The same goes for musical notation, which generally functions as a command to produce sound and therefore records it only very incompletely.[94] Studies focusing on the history of the transcription of the spoken word have shown that the first half of the nineteenth century saw the rise of discontentment with the use of written language as a way to represent speech. Speakers in parliament, as well as journalists, increasingly showed anxiety over the accuracy of written reports in representing their political speeches, for example, leading some commentators to exclaim that "a phonograph" would be needed to "catch the accent" of particular speakers.[95] I would argue that the shift toward a more material understanding of the voice itself also played a role in such anxieties, and that the drive toward accurate visual representations of sound and those of the vocal organs were connected—at least for those who had a professional interest in the human voice.

Accuracy in visual representation was therefore not just an issue in transcription, but became highly important across the board: the vocal organ was to be depicted in detail as well. Scientists had been doing that for a while, of course, as anatomical drawings were available in the early modern era, but the nineteenth century saw a rise in interest in including such drawings in works directed at amateurs or dealing with issues outside the medical realm. Manuals for singing as well as self-help books would include equally careful (if perhaps less lengthy) explanations of the mechanics of the voice as medical treatises. They often included the scientific terms "larynx" and "pharynx," and explicitly cleared common misconceptions about the vocal mechanism (such as the character of the "false" vocal cords, for example). Anatomical depictions of the larynx were didactic and lent themselves above all to the comparative work that would arise with the increasing popularity of statistics: for those to whom tables and numbers did not make immediate sense, drawings of the "average" male or female larynx would be much more elucidating. Anatomical drawings were used throughout the nineteenth century (and beyond), but by the second half of the century were thought to have one major drawback: they showed a "dead" larynx, one divorced from the human body that made it work and—more importantly—from the movement that actually constituted vocal practice. Unless an image of a cross section was included as well, these images did not show the mechanism of the voice at all, because that was largely located within the larynx.

Experts were at pains to explain to their audience that the voice was not produced in some sort of mechanical box, but, since the working of the voice remained hidden, the metaphor of the voice box remained current as well . . . until the laryngoscope provided them with a new way to visualize the voice. The new invention not only allowed its users to see the hidden mechanism of the voice at work, it also made new ways of depicting their knowledge possible. Almost immediately after the invention of the laryngoscope, laryngoscopic views of the voice were included in textbooks. This allowed medical experts to compare their patients' larynxes to a range of depicted pathological ones, but also included attempts to distinguish between correct (i.e., healthy and beautiful) uses of the voice and wrong ones. Like the anatomical drawing, laryngoscopic views would make their way into manuals for song and speech as well, where they were used less as diagnostic tools and more as encouragement toward correct vocal performance. In 1861, depictions of larynxes based on laryngoscopy were included in a paper on phonetics, showing not only how easily the instrument traveled between disciplines, but also the level of detail (and the level of trust) that was assigned to it. This paper was one of many in which it was assumed that things like tessitura, or even a particular sung note, could be distinguished by visual observation.[96]

The most important advantage these depictions had was, of course, that they showed larynxes doing rather than being something. Anatomical depictions could show some aspect of biologized identity (e.g., the female and male larynx in Fournié's work indicated something about gender and its associated vocal sounds). Laryngoscopic examinations and their resulting drawings could show the soprano's high notes, or the hoarse man's croak, in the making.

Attempts to better visualize both vocal sound and the mechanism of the voice would continue throughout the nineteenth century. Inventions such as the gramophone and the telephone, which enabled acoustic reproduction of the voice, caught the public imagination and became the great commercial successes we still know them to be. However, at the time of their invention they did not dampen inventors' ambitions to create ever new ways to turn the voice into something that could be archived and, above all, observed. At the time when the laryngoscope was introduced, the possibility of sound recording and replay existed only in the imagination, and the eerie experience of hearing a human voice coming out of an electronic device rather than a human body was unavailable. The opposite had happened: with the introduction of the laryngoscope, definitions of the voice had slightly shifted. When mid-nineteenth-century speakers (or at least vocal experts) talked about the voice, they no longer referred to a sound they recognized as song or speech and could trace back to a throat; they meant the vocal organs themselves. For users of the laryngoscope, what they saw by using the new invention was understood as the voice itself. And thus, voices could be ill or healthy in their own right (rather than merely signal pathology or health in their owners' bodies), they could be young or old, male or female, or rather they could perform all these characteristics and this physical performance could be seen as well as heard. Around 1855, the voice became a material, tangible *something* rather than the uncanny and invisible emanation of the soul it had been for so long.

INTERMEZZO 1

The Laryngoscope and Its Many Fathers

> It is a well-known fact that the first experimenter who succeeded in obtaining a view of his own larynx is a distinguished professor of music in this town, M. Garcia. . . . He attained the desired object by a very simple plan. Standing with his back to the sun, he held a looking-glass in his left hand before his face, the sun's rays were thus reflected by the glass into his open mouth. When he introduced a dentist's mirror, previously warmed, into the back of his mouth, and thus he saw the reflection of his larynx in the looking-glass.
>
> —George Johnson

And so, with this surprisingly simple action, the laryngoscope was born. Manuel Garcia reportedly conducted his first experiments in 1854 and gave a paper entitled "Physiological Observation on the Human Voice" at the Royal Society of Medicine. The search for a convenient instrument for observing the larynx at work had been going on for almost a century by then, with several physicians experimenting with mirrors and specula, but in the end a singing master (brother of the famous Maria Malibran and Pauline Viardot) turned out to be the first person in the world to see a living human voice: his own.

"Garcia's claim to originality in the matter of autolaryngoscopy" was generally seen to be "quite distinct and indisputable."[1] Nevertheless, the invention of laryngoscopy as a practice was the work of many hands, and it soon became clear that Garcia's work as an inventor was only a small part of it. According to James Prosser, who offered a history of laryngoscopy in his 1873 *Lessons in Laryngoscopy*, the story of the instrument was less of a singular flash of brilliance, and more of a conflagration of events across Europe. The invention of the laryngoscope, he stated,

> closely resembles that of many other inventions. We find that several persons at various intervals were intent on the endeavour to obtain a view of the interior of the larynx during life. Some, indeed, seem to have actually seen the glottis, others, without clearly defining them, discerned in a mirror parts otherwise invisible, and all succeeded more or less, according to the patience they brought to bear on the problem, and the methods and instruments they employed. Thus it is that several persons only just missed being recognised as the inventors of the laryngoscope, and when at length the instrument was perfected and its value demonstrated, more than one claimed a share in the honour of discovering so great an addition to our means of investigating and treating disease.[2]

He mentions Green, Czermak, and Türck as major contenders for the inventor's title, but finally more or less settles on Manuel Garcia—as most of his contemporaries and subsequent historians have done.[3] Authors of manuals for laryngoscopy such as Morrell Mackenzie and George Johnson agreed that, as Archibald Garrod noted quite simply in his 1886 *Introduction to the Use of the Laryngoscope*, the instrument "was invented by M. Manuel Garcia in 1855."[4] Malcolm Sterling Mackinlay, in his 1908 biography of Garcia, noted that "It was in 1854 (the year which saw the ultimatum of England and France presented to St Petersburg, the prelude to the Crimean War) that the important invention was made—or, as the maestro with characteristic modesty described it, 'the idea dawned on him'—of the laryngoscope." In a meticulously drawn chronology of ideas, events, lectures, and publications, the biographer legitimates this statement, carefully including Ludwig Türck's and Johann Czermak's (both Austrian-German) ever so slightly later involvement with laryngoscopy.[5] Meanwhile, Türck vociferously claimed he himself was the first to recognize the value of the laryngoscope to the physician,[6] although others named Czermak as the "resuscitator and reviver" of the instrument and therefore the "father of laryngoscopy."[7] Morrell Mackenzie, who would soon become a leading authority on laryngoscopy and founder of a Hospital for Diseases of the Throat, diplomatically afforded a place to all three in his history of the laryngoscope, alongside various other, earlier experiments in visualizing the vocal cords.[8]

In the preface to his *The Use of the Laryngoscope*, Mackenzie told the story of visiting Professor Czermak in Pesth in 1859 to learn about laryngoscopy "whilst the art was still in its cradle."[9] It is clear that, for Mackenzie at least, the people standing around that cradle as potential fathers or godfathers were many, and constituted a diverse group. Above all, the invention of the laryngoscope appears in his manual as a thoroughly international and interdisciplinary endeavor—and as a long process.

"It may seem strange to some that it was not till the middle of the last century that an instrument was invented for examining the lower part of the throat during life, nor till more than a hundred years later that that instrument was sufficiently improved and simplified to be capable of general application."[10] The first instrument he referred to was an invention of the French *accoucheur* (ob-gyn) M. Levret, whose work heads a narrative of international inventing-tag, with the story traveling through Germany, Switzerland, England, and France again before 1854, when "the idea of employing mirrors for studying the interior of the larynx during singing occurred to M. Manuel Garcia."[11] Garcia was identified not only as a "long distinguished singing-master in London," but also as "a Frenchman by birth, and a Spaniard by descent."[12] Exactly the kind of mobile, international figure that—in Mackenzie's narrative—seemed destined to play an important role in the history of the instrument that would soon be developed more thoroughly in Vienna, Pesth, and London. Moreover, as a singer and singing master, Garcia had been particularly adept at controlling his own throat, which helped him in his autolaryngoscopic experiments. This reliance on self-observation remained important throughout the first years of the development of laryngoscopy as a medical practice as well. Czermak, whom Mackenzie credited with "superior genius, untiring perseverance and natural dexterity," would be the one to create "the art of laryngoscopy."[13] His natural dexterity was not only a matter of medical skill, however. According to Mackenzie, "Czermak's investigations were at first confined to his own larynx, and his success must in part be attributed to his great physical advantages. Possessed of a most capacious pharynx, small tonsils and uvula, and a large laryngeal aperture, it would be difficult to find a subject better suited for laryngoscopy."[14] Autolaryngoscopy would remain an important part of laryngoscopic practice for decades: in medical manuals, students were encouraged to first practice on their own throats. This would not only give them a chance to develop sufficient skill with the instrument (and potentially empathy for the patient in light of the discomfort of the procedure), it would also acquaint them with the sight of a normal, healthy voice—assuming they possessed one, of course. If the number of published manuals is any indication, aspiring physicians all around Europe must have been examining their own throats in the second half of the nineteenth century. Meanwhile, Morell Mackenzie developed his skill, and his reputation as an expert on the pathology but also the visualization of the voice. The Wellcome collection holds fragments of his professional and personal archive, which shows that in the 1870s the Mackenzies were entertaining a visiting Professor Czermak in their London home, and that they were also in regular contact with Garcia. The Hospital for Diseases of the Throat was a local one, but Mackenzie was clearly keen to connect his work to the international network of inventors and experts he considered to be the main context of his work and audience for his publications. Around 1890 he managed to produce photographs of some of his cases, thus providing him with the means to show the living voice not only to medical students, but also to the international community of experts he had been guiding around his hospital for decades.

In doing so, he took the laryngoscope far beyond how its inventor had put it to use. When Sterling Mackinlay published the glowing biography of his father-in-law, "being a memoir of Manuel Garcia's life and labours for the advancement of music and science," the laryngoscope occupied an important but rather small part of the narrative. According to the biographer, the instrument "connected entirely with his work as a teacher of singing."[15] Garcia had no interest in vocal health on a general or hygienic scale; he merely wanted to "see a healthy glottis in the very act of singing" so he could confirm the theories on singing and voice production on which he had founded his famous singing method. As soon as he had achieved that goal—and as soon as he had published his findings—he ceased to be interested.[16] Moreover, as the biographer noted, it took several years before Türck and Czermak picked up his paper and saw the scientific and medical potential of the instrument, "as so often happens."[17] Although his biographer presents Garcia as a great innovator, responsible for a revolution in medicine which Mackinlay asserted had benefited "3 percent of the entire human race,"[18] he depicts Garcia as a musician and teacher first. The biography is far more concerned with Garcia's famous pupils than it is with his invention, and even in the short section that deals with the laryngoscope explicitly, Garcia is represented as a "discoverer of the hidden land" rather than its cartographer.[19]

The story told in the biography is not one of careful preparation and scientific precision, but rather a tale of someone elegantly and almost magically stumbling upon the exact answers to his one wish: "if only I could see the glottis."[20]

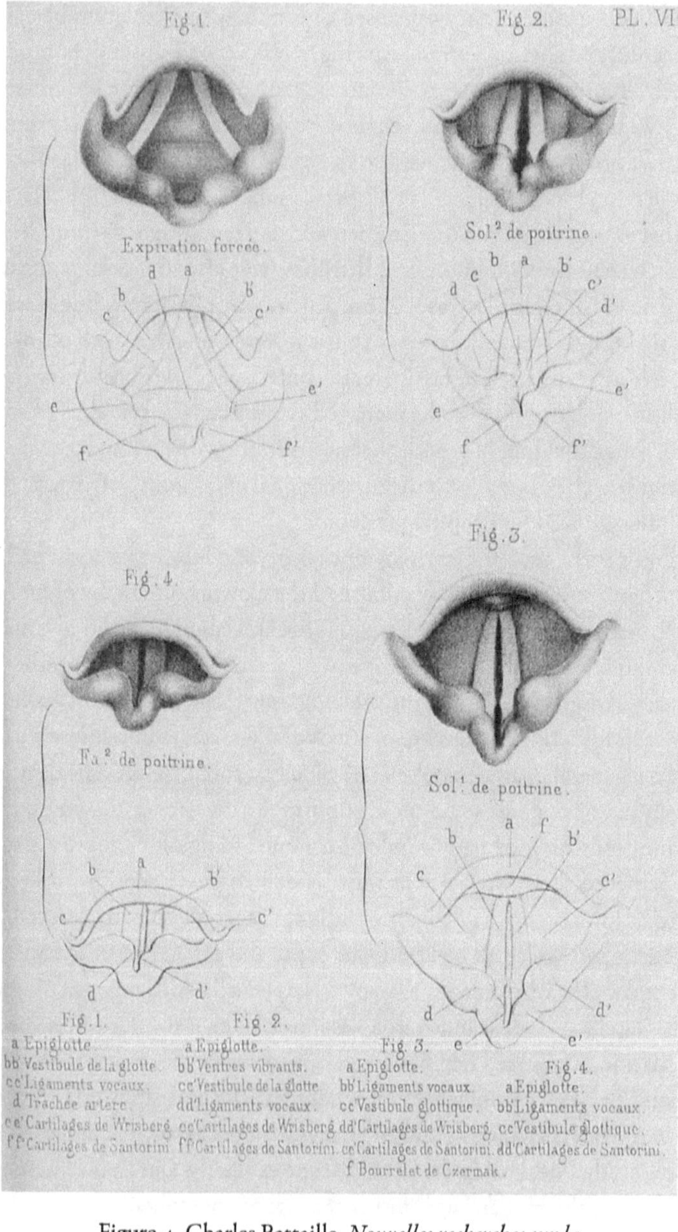

Figure 4. Charles Battaille, *Nouvelles recherches sur la phonation: Mémoire présenté et lu à l'académie des sciences le 15 avril 1861*, Paris: Victor Masson et fils, 1861. Pl. VI, Fig. 3. Courtesy of the Royal College of Surgeons of England.

The discovery of the laryngoscope thus appears here as a succession of brilliant flashes and lucky coincidences which, for later scientific and medical use, had to be painstakingly recreated—for example by developing a method that allowed the use of artificial light. Nevertheless, the basic characteristics of the instrument remained the same, and numerous manuals would be published to teach students to insert the mirror at the angle Garcia had reportedly so easily found.

The history of the laryngoscope's invention remains a site of (moderate) contention, but Garcia's role as the poster boy for a revolutionary turn toward the study of the living voice and the subsequent scientization of vocal pedagogy in the second half of the nineteenth century is uncontested. More than to the actual empirical results of his research or even to his experimental methods, he owes this reputation to his capacity to bridge the divide between the two most active disciplines concerned with vocal practice: anatomy and song. Garcia is cited not only in scientific studies and medical manuals for his findings, but also—and perhaps even more so—in manuals for teachers and students of singing technique. Singers and singing teachers discussed Garcia's views on the registers of the voice. Many were scandalized by his theory of the "shock of the glottis," a somewhat violent way to express a technique that allows the singer to initiate a tone without added aspiration.[21] Garcia always claimed his notion of this "vocal attack" was misunderstood by many of his readers, and his responses to his detractors prolonged the discussion. His most important contribution to the craft, however, was that he convinced many of his contemporary artists and educators of the need to include physiological knowledge and, most notably, the results of laryngoscopic research in their vocal training.

As a singer and voice teacher, Garcia could boast an impressive pedigree. His father, tenor Manuel Garcia Senior, sang in Paris, London, and New York and was taught by Giovanni Ansini. Through Ansini, the Garcias (including Maria Malibran and Pauline Viardot) could trace their vocal training back to none other than the famous Nicola Porpora. Porpora was active as a composer and singing teacher in the early eighteenth century, training Farinelli and other castratos. The connection to this artistic lineage was a valuable one in the nineteenth century, when the old Italian masters had receded far enough into memory to have turned into almost mythical creatures, representing a great art (almost) lost to mankind. By connecting their own modern methods to this celebrated history of craftsmanship, the Garcias managed to forge a formidable reputation as a singing dynasty. The younger Manuel Garcia's own contribution to vocal artistry was widely acknowledged, owing much to the success of his numerous pupils: he taught both at the Conservatoire de Paris and the Royal Academy of

Music, training among others Jenny Lind, Henry Wood, and Christina Nilsson. His legacy as a vocal teacher was later carried on by other influential teachers such as Mathilde Marchesi, and thus remained current well into the twentieth century (Nellie Melba was educated by former pupils of Garcia, for example). In her autobiography, Marchesi dwelt on "Garcia and his invention," noting that "his ideas on the female voice and its development were a revelation to" her, and "the foundation of [her] own future career."[22]

CHAPTER 2

Acquiring a Voice

Elocution and Education

The energy and attention that went into attempts to see—and by those means know—the voice were both driven and complicated by shifting cultural ideas about the nature of the human voice. As a product of the body, it appeared both as a given (a gift of nature, for some, or from God, for many) and as the result of human effort and practice. Like many other aspects of physical appearance and achievement, vocal skills, pathologies, and excellence were ascribed to a curious combination of assumptions about morality and health. Developing a beautiful, healthy voice was therefore understood as both letting natural growth take its intrinsically perfect course, *and* cultivating vocality through practices of song, elocution, gymnastics, and so on, in order to remove pathological sounds and to encourage those heard as healthy and proper. It was, in other words, not just the laryngoscope or the rise in visualization or physiology that made the voice material in the nineteenth century, it was also a broad cultural instrumentarium that increasingly encouraged vocalists to consciously engage with their voices as a product of both artifice and nature, and to turn to medical, therapeutic, and biological knowledge to cultivate what nature had bestowed upon them.

This chapter focuses on the practical and performative aspects of such an ambiguously cultivated voice. As most therapists, canto teachers, and scientists were insistent that their methods did, curiously, produce a natural voice, I avoid drawing sharp distinctions between the natural and the artificial in my analysis of their work. This chapter therefore centers vocal propriety, instead of vocal cultivation per se. Although propriety was an important, and highly charged, issue for nineteenth-century historical actors, the term is not an established category of analysis. However, the term carries implications of class difference, and particularly of middle-class sensibilities, that are central to many of the practices this chapter is concerned with, without being necessarily identified as such. Additionally, propriety crosses some of the boundaries that often define research on social behavior and cultural practices in the nineteenth century. Research on what could roughly be called middle-class cultural repertoires of morality and social distinction tends to be focused on particular nations or linguistic contexts. For Victorian Britain, the focus is often on the connection between manners and morals,[1] for the nineteenth-century German lands, the concept of *Bürgerlichkeit* has been defining of the field,[2] and for France this type of historiography has been conceptualized as a history of sensibility.[3] All three of these traditions have been called "new social history" of one type or another—that is, social histories that include cultural beliefs and representations. They also all have developed further toward studies of embodied practices that are implicitly connected to class, and often explicitly gendered, ethnicized, and modulated by age. These social histories of bodies and daily practice are particularly useful for thinking through "performativity" in the context of universalist nineteenth-century beliefs about beauty and health.

The concept of vocal propriety I will mobilize in this chapter, that is, the implicit and explicit modes of social distinction that were articulated through vocal practice, has been influenced mostly by these newer iterations of new social history, such as Martin Hewitt's conceptualization of "social knowledge," Ute Frevert's histories of emotion, and Delphine Gardey's histories of embodied practice.[4] The concept consciously engages with longer traditions of social history in these three linguistic and geopolitical contexts to create room for the transnational history this book is tracing, in which overlap and transnational exchanges and continuities tend to outweigh national differences. Vocal propriety, in other words, is approached in the following pages as a framework within which linguistically and geographically distinct and changing vocal practices can be contextualized at the interface of cultural repertoire and social distinction.

Developing vocal propriety, or acquiring a proper voice at these porous interfaces of culture, society, and biology, appeared as a precarious balancing act between trusting and respecting what was given and bending the body to one's will. Tethered to both ideologies of an ideal, god-given nature and increasingly common ideas about a world governed by natural laws, the vocalist seeking to develop a voice fit for society or art walked a narrow path. This was not entirely new, of course. As Jennifer Richards has shown, early modern ideas about vocal practice were dependent on visualizations of the voice (albeit as anatomical drawings, rather than imagery derived from a laryngeal mirror), and also presented a voice that was "made out of air" that flowed through the organs of speech.[5] "Lips, teeth, tongue, palate and throat" figured as the "stops that make the music of the human voice." Meanwhile, various practices of vocal cultivation—including those of the by then generally discounted "artificial" style of the baroque—also continued to circulate. Rather than a simple move toward a modern science based on observation and emerging notions of objectivity, vocal knowledge in the first half of the nineteenth century developed as a site of constant tension between images of the voice as a physical object made up of organic tissue, and the voice as a site of artful cultivation.

In both iterations, the ambiguous interstices of having and acquiring a voice would come to be connected to narratives of evolution and be expressed in vocabularies of change at the species level and in stories of civilization. In the first half of the nineteenth century, however, the stories of a perfectible nature, next to those of an ideal state of nature created by an infallible god, were inflected by human voices whose training had to be imagined in different terms. For English-speaking members of the British empire, for example, elocution became an important aspect of the civilizing mission that they had assigned themselves, and the standardization of language (which served, among other things, as a way to eradicate other modes of speech) went hand in hand with a standardization of pronunciation and sound.[6] The cultivation of a god-given, naturally perfect voice—as if tending to the garden of Eden—required subtle management and a rich vocabulary to think around change in the context of a biology that was generally seen as becoming more and more fixed.[7] It is no wonder, perhaps, that children and their malleable bodies and voices would provide such a fertile ground for the development of new ideas about the natural human voice throughout the nineteenth century.[8] While images and observations *of* children were often central to the development of the cultivated voice, the observing, describing, and systematic cultivating was largely carried out by adults with multiple claims to authority—such as, for example, choir master and mass singing proponent Joseph Mainzer. Mainzer assured his readers that the best time to develop the voice was in early childhood, "at a time when, ultimately, these organs are still flexible and open to all impressions."[9] Similarly, a pamphlet by an anonymous physician, published in 1843, exhorted readers to think of vocal training as an endeavor involving the whole body, to be undertaken in youth, and requiring a long-term approach: "moderate and well-devised gymnastic exercises, if resorted to in early youth, would evidently produce good effects, by promoting the development of the chest, and the healthy action of the lungs."[10]

To come to vocal perfection, according to this combined perspective on natural gifts and cultural practice, one needed resilience, endurance, and constant exercise. One also needed time. The perfect natural voice was not simply given, but carefully cultivated and tended to over the years—and especially young voices were understood to require particular care and protection from overexertion, strain, and unhealthy habits. Moreover, by virtue of being emitted by a young, unpracticed body (or by an overstrained middle-aged one), the voice itself was heard to have age-related characteristics and issues. In fact, because practice played such a role in shaping the natural voice, a host of other "natural" contextual and physical factors played into the making of both the speaking and singing voice. Gender would be vocalized in particular ways, sonified by larynxes understood as male, female, or infantile. So would different social and geographical features, as different functions in society, different ways to interact with the world, and different material environments would affect the way the voice was developed and therefore affect the vocal organs and the voice itself. The manifold practices, therapies, "cures," and exercises developed by experts in this period made the voice and its material characteristics into a social object.

In this chapter, I trace the different ways in which vocal propriety was achieved and maintained, and the different modes of vocalizing identities modulated by social class and profession, age, and gender (roughly in that order). While the disciplinary tussles and scientific formalization of knowledge on the voice described in the previous chapter occupied only a small group of experts, the kind of practices they prescribed and described, as well as the new ways of imagining and understanding voices they developed, could have a far greater reach. Moreover, as the professional fields occupied with the voice grew, it became increasingly common to think of vocal "trouble" and health as the domain of medicine and therapy. Vocal propriety, a somewhat wider concept than vocal beauty, would no longer be the sole purview of the singing master, as the natural physical voice was understood to be beautiful through health and its cultivation, rather than through art and artifice. In the next sections, I therefore focus on the confluence of medical and social discourses of vocal pathology and health, and the importance of ideas on the natural development of vocal propriety throughout one's lifetime, but I also discuss the continuing importance of vocal distinction (between differently gendered voices, but also between well-developed and less well-developed voices) and musical traditions of vocal education and excellence. In the acquisition of vocal propriety, the materiality of the voice and its performative nature worked consistently in tandem, across different fields of practice and expertise.

Propriety and Pathology

The work of producing vocal propriety was assigned to every vocalizer, but also, and especially, to modern singing masters and speech therapists. Both were increasingly imagined as experts in vocal physiology, and tasked with "fixing" the fickle features of individual vocal habits into proper vocal practices. Acquiring a proper voice, or helping somebody else do so, consisted of making stable what appeared as too fluid, too changeable, and too uncertain for comfort. Through educational as well as medical practices, patients and students together with physicians and teachers defined what constituted a male or female larynx, what could count as healthy or unhealthy vocalization, which sounds were pleasant and which ones were not; in other words, they established what constituted a proper voice at the interface of social expectations, cultural ideals, and medical norms. Their definitions also allowed them to devise cures, or means of assimilation, for those voices that diverged from what they considered proper. Addressing a (largely) middle-class audience, and often hailing from middle-class backgrounds themselves, these newfangled experts produced scientific insights that—regardless of their methods or degrees—tended to reproduce middle-class expectations through the vocabularies of health, nature, and normality. Pathologizing improper voices and subsequently fixing what thus seemed broken buttressed those ideas.

The fluid nature of the boundaries between pathology and impropriety was clear to most experts and practitioners—and was one of the problems that their works intended to solve. Large numbers of publications delved into questions like whether a stammer was a "habit" or a "disease." Or they addressed such questions more directly by asking whether an affliction could be cured. The frequency with which phoniatrists and laryngologists referred to the value of a good singing master shows to what extent these so-called cures dovetailed with methods to acquire grace. In 1898, a slim volume was published entitled *The Voice: Or the Physiologist Versus the Singing Master,* in which medical and musical methods were presented next to each other, strengthening the well-established idea that medical and musical modes of vocal practice were closely related.[11] Portraying methods of training the voice as a cure, however, subtly shifted the social implications of learning how to speak in polite society. Whereas manuals on polite conversation, riding, playing the piano, and so on proliferated throughout the early modern period and circulated among the higher classes,[12] etiquette manuals in the nineteenth century would increasingly be seen as a sign of a desire for social climbing—and thus signal one's lowly background. Manuals for good speech couched in scientific terms, it seems, cleverly avoided this implication by suggesting that the reader of such work *was,* in fact, innately capable of middle-class propriety (presented as a norm defined by nature), but was prevented from displaying it because of a curable illness.

Conversely, these entanglements of middle-class norms and medical ideas about health resulted in the pathologization of vocalizations that diverged from the norm. As the (upper) middle class came to set the norm, speakers in both the lower and higher social segments of society were heard as deviant. Upper-class inflections were, in some contexts at least, starting to sound suspect: as norms of public power shifted, so did practices of articulating it. In contexts like democratic representation and jurisdiction, places where the new middle-class ideas of power and governmentality reigned, new norms around rational speech and authoritative masculinity were forged. Speakers who adhered to older modes of public speech gradually lost influence and respect. This seems to have happened, for example, to Edward Bulwer-Lytton, whose flowery mode of delivery in the House of Commons was roundly ridiculed as out of date because it contained "sparkling gems, much ingenious antithesis, and some effected and carefully polished declamation. It is unnecessary to say that a style so artificial and language so ornate are never found in extemporaneous oratory."[13] Mostly, however, criticism of speech that departed from the middle-class norm was directed at the lower classes. The voices of those who made their living in the open air were often described as merely unpleasant—rough, ugly, but also strong and healthy. Even the cries of hawkers and peddlers in the city, according to William Gardiner, were not without charm: "The articles of commerce being chiefly brought from the country, were cried in the artless tones of the peasantry, founded upon the natural exclamations of the voice."[14]

The voices of the industrial working classes, however, even though their timbre was often described in similar ways, became the object of great scrutiny and research. The rough quality of the voices of the laborer or the prostitute was ascribed to the unsanitary conditions in which they sounded, the unhealthy state of their bodies and minds, and also to their practices of speech. The "hoarse, discordant" sounds of syphilitic voices, or throats marred by alcoholism, were particularly reviled.[15] In a sense, the pathologization of the lower-class voice was merely the mirror image of the therapeutic discourses surrounding cures for faulty voice production: it was predicated upon the notion that the voice, as a corporeal feature, was both naturally given and "made" through practice. And the vocal practice of the lower classes was, according to normative frameworks of standardized language and rational tone, inherently wrong. Or rather, nonnormative speech was both pathologized and increasingly ascribed to a particular class of speakers. French playwright Édouard Mennechet, for example, identified both accents and dialects as "vices de prononciation" and "vice de la parole" on a par with rhotacisms. They adulterated speech like lime penetrating the soil.[16]

Likewise, the sounds produced by deaf persons were often heard as coarse and uncivilized—a quality attributed to both their perceived lack of natural ability and assumptions about their social backgrounds and upbringing. Although educators of deaf children were at pains to show that their charges had religious feelings and understandings (and therefore possessed a soul), their intelligence was very much in question. Sign language was generally understood to be a far less sophisticated system of communication that did not allow its speakers to go beyond the concrete. A memoir written by educator William Sleight, detailing the life story of a deceased pupil, gives us an insight into how insistence on the ability of young deaf people to learn language sat uneasily beside the conviction that deafness impeded intelligence.[17] In the memoir, the young boy is described as a "thinking boy," with a keen interest in spiritual life (and a particular affinity for the Bible), but also almost like a wild child, from humble circumstances, with little or no education and an unwieldy temper—something that was presented as common in deaf children.

> The Deaf and Dumb have signs of their own, by which they could easily make themselves understood, on all common subjects, if those among whom they are placed would give them their attention. It too often happens however, that by a shake of the head, or a wave of the hand, the child is put on one side without noticing what he has to say; and thus his temper is irritated and spoiled, and as he grows older he becomes jealous and passionate. Hence the character so universally given to the Deaf and Dumb. But when he is placed early under instruction, he is as easily brought into discipline and order as another child, and no perceivable difference is exhibited in his temper.[18]

Despite the communicative ability of signers, the author still stressed that signs were "a poor substitute" for spoken language and further education would be required to make them comply with the rules of society and act with propriety.[19]

The interplay between natural vocal faults and acquired habits served to define norms of impropriety similarly to the ways in which vocal training and ideas of natural talent converged to construct propriety and normality in the case of speakers and singers that were heard as exemplary, as I show later in this chapter. Social norms and expectations played a large part in that process, as power imbalances served to co-define health and propriety in the first place. A good match between social identity and its articulation therefore lay at the heart of processes of normatization—and a mismatch was read as an anomaly in need of a cure. Women with very low-pitched voices or men producing high, flute-y sounds were presented as such mismatches between voice and identity.

Such a mismatch was heard as pathology, but that made it by no means stable or permanent. The making of the adult gendered voice was imagined as a fraught process of both natural growth and the careful, learned acquisition of gendered characteristics. This was perhaps the most obvious in how an important aspect of vocalization—breathing—was gendered. A number of authors, particularly those with a more therapeutic pedagogical approach, voiced concerns over the restrictions fashion could place on, especially, women's breathing. If, as many of these authors insisted, the lungs were an integral part of the vocal apparatus, they too had to be trained carefully so that they could properly fulfill their natural function. In stammering pamphlets, singing courses, and rhetorical manuals alike, the art of breathing was given significant space and attention, thus further cementing the more holistic approach to the voice put forward by physiologists, and also consolidating a narrative centered on a nonintuitive, perfectible nature ascribed to the normative expectations of the human voice. Corsets were often identified as yet another piece of early modern artifice unsuited to the modern age that could have a "pernicious influence" on the respiratory system.[20] Tight-lacing, it was argued, prevented women from proper diaphragmatic breathing, thereby also limiting their ability to produce voice. "All belts, corsets and tight clothing is noxious to exercising and to health in general," one manual, *The Art of Breathing for Song, Speech and Life*, stated confidently.[21] The arguments made around gendered breathing rarely followed a clear causal path, however. While the restrictive nature of corsets was criticized widely, there also seems to have been a strong belief that women's clavicular breathing was a natural sign of femininity as well.[22] Although men, too, were targeted by various exercises and instructions to develop their lungs and techniques to breathe naturally, their assumed superior strength was generally identified as a sign of superior breathing as well. In other words, vocal articulations of identity markers, like gender, were always heard at the crossroads of nature and practice.

Discourses around the corset and healthy breathing were further complicated by the fuzzy meaning of the terminology of what is natural, both historically and now. A persistent confusion between the natural as intuitive and the natural as biological truth runs through any discussion of vocal practice—including in manuals for song and speech published in the eighteenth or the twenty-first century. The nineteenth century, however, was particularly insistent on both glorifying intuition in theory and distrusting it in practice. Imagining modern man—in his newly industrialized society—as a wholly new being seems to have led many scientific observers to the conclusion that modern civilization had deprived the European, white (male) human adult of all connections to his supposedly natural urges. Overcivilized and therefore nervous, these creatures developed speech impediments, forgot how to move their bodies and use their organs properly, and, therefore, were forced to systematically study the function and nature of the human body in order to develop sophisticated methods to reintegrate natural movements and practices into their daily lives. This may be one reason why creators of stammering cures were so intent on pointing out that various groups and populations they identified as less developed or less civilized did not stammer. For the middle classes, for whom stammering was understood to be a common pathology as a result of their overcivilization, numerous curative practices and exercises were devised. In practice, many of these borrowed from older techniques used in the training of singers, but their authors underpinned them with ideas about physiology and the mechanism of breathing. Much like vocal training for the stage, the training of the healthy, fluent voice became a profession in its own right, and one that fit with the rise of other modern professions that straddled the divide between paid work and the sort of intellectual pursuits previously reserved for gentlemen.

The new approach to voice training—which diverged from older methods that taught song as craftmanship rather than as a matter of health and well-being—also coincided with a similar shift in attitudes about work: in arenas of production and trade, too, craftmanship and seasonal labor were slowly giving way to a model of a professional class and a working class. Both classes depended on their natural gifts of, respectively, intelligence and physical strength, while education and physical exercise gained importance for developing those gifts, particularly in those not so favored by nature.[23] The voice was therefore approached much like visible musculature: as something that a particular class of people (and particularly middle-class men) lacked, but that they could develop through exercise. James Mackness, focusing on the illnesses caused by public speech, thus mused on the reasons why clergymen were so much more likely to be "afflicted" with hoarseness than men in other professions necessitating lengthy speeches: "We need not be surprised . . . when we reflect by what circumstances the choice of the clerical profession is sometimes determined. A lad perhaps is delicate in health, and is hence led in early life to delight in quiet and sedentary occupations, seeking his amusement rather in solitary reading than in the noisy but invigorating sports of his more robust companions."[24] And so, like his sedentary counterpart in the professions, the parson needed to be encouraged to strengthen his body, not for its own sake, but in order to underpin the health of his voice:

> there must be first, daily exercise in the open air, either on foot or on horseback. As the opinions of society, and their own delicate, perhaps even scrupulous sense of propriety, preclude the younger clergy from following those field-sports which often serve to give robustness to other men of study, it is the more necessary that they should avail themselves to the utmost of such out-door exercises as are still permitted them. The most important, and at once the most accessible of these, is walking, than which nothing is perhaps more conducive to health. This should be persevered in without over-fastidiousness as to the state of the weather.[25]

Although women did, of course, speak professionally too—as teachers, for example—their speech was far more often described as musical and the result of natural charm.[26] To some degree, this notion of women as creatures of nature carried over even to the world of professional vocalists, whose voices were likened to those of birds and angels, but rarely scrutinized as the product of professional training. On stage, the rise of female stars of operatic art would lead to a veneration of some female singers that elevated them to god-like heights, where their vocal powers appeared to be nearly magical. To be a star on the operatic stage, a man would have to be a hero, intrepidly disciplining and strengthening his body. A woman would be a diva, a creature of caprice, wild emotion, and natural talent. This understanding of the female embodied voice as somehow less controlled than that of adult men also dovetailed with gendered discourses about puberty and maturation.

Youth and Education

The topics of maturation, growth, and physical development took up a central place in discourses on vocal education. Because a "natural" voice production was largely understood to be intrinsically more healthy and therefore more beautiful and proper, voices that could be heard as unfettered by the intrusions of the bad habits associated with modern urban life (like the corset mentioned above) appeared as ideals. Experts therefore posited that voices were at their best in young bodies, and that the development of a good healthy voice was a matter of protecting this naturally proper instrument. As the *Gazette Musicale de Paris* suggested in 1835, composers did not write for old voices, because those lacked the luster of youth: "What repertoire would you find, I ask, of roles written for—I will not say worn out voices, no reasonable composer would entertain such ideas—but for voices that are past the height of their sonority? Musical passion is all exterior, all fire, all exaggeration. It does not argue, it acts. To act, you have to be young, musical passion cannot do without youth."[27] Children, in particular, became a central cause for concern not only to specialists interested in vocal health, but to a wider audience too. They were presented as a means to further the health of the nation through vocal education. In 1839, four years after their piece on youthful *cantatrices*, the *Gazette Musicale de Paris* ran an article by the mostly political journalist Adolphe Gueroult on singing for schoolchildren. In it, Gueroult imagines a reactionary older woman, versed in a culture in which music was the preserve of the higher classes and who therefore fails to see the value of song for the populace: "the mere idea, she said, to teach them music! Sure, they will need that to keep their kitchen and to clean their children! It will be a nursery for actresses. *Luckily* they are too ugly for such a thing!"[28] Set up to symbolize an older—and particularly an older French—view on both music and the education of the lower classes, this grand lady on her balcony served to underline the new, modern message the author of the article aimed to spread: in the healthy modern nation (exemplified by Germany and Italy), *all* children contributed to a healthy future of the population, and musical training played an important part in this project. Gueroult continues: "And why, madame, would your cook not sing whilst stirring your pots! Would it be so bad, if instead of singing sharp, she'd sing in tune? . . . let the people learn music and believe me, you will thank those who teach them how to sing, because nothing lifts the spirit like singing."[29] The expression of belonging and community through song had a long history, of course, but in the nineteenth century these ideas about belonging started to be organized and ethnicized in new ways. The first half of the century saw the establishment or new independence of a number of European nations. These modern entities—which historians have understood as consisting of imagined communities and founded on invented traditions—were also discursively provided with biological underpinnings. Early nationalists argued that the people belonging to these nations were a naturally cohesive group. The (collective) singing particularly of children and young people could easily be imagined as part of the natural bonding rituals of such groups. Choir master Laurent de Rillé noted in 1856 that "primitive" peoples expressed their emotions through song, and that therefore collective singing provided the means for the nation—which was a mere extension or development of primitive humanity—to express itself. "When the whole tribe, when the nation is moved, the tribe, the nation, assembles and millions of unified voices intone a religious or warriors' hymn."[30] The future of the nation—or the tribe, as de Rillé would have it—was put in the mouths and throats of its young members. This required the rigorous training of all would-be citizens, for "The real purpose of choral societies is to teach their members orderly habits and to make them accept and practice the principle of voluntary solidarity; to spread the taste for popular music, the germs for musical instruction, throughout the country."[31] It also required practices of exclusion—which historians have shown to be central to the development of nation-states and democracy—or the application of seemingly natural standards: "It is necessary in a choir, as it is in a besieged town, to eliminate all superfluous mouths."[32] The tension present in de Rillé's discourse on song was present more broadly in narratives on vocal education, propriety, and health: presented as an issue of public health, vocal normality was required of all, and therefore resulted both in broad inclusive policies to educate or cure the many, and in disciplinary measures to remove the faults (or, if necessary, the sheer presence) of the few. In the case of children, and lower-class children in particular, this created a rather tricky conundrum for educators of a modern, biologistic bent. Whereas the accents, rough tones, and uncultured speech of the lower classes presented endless possibilities for a civilizing mission, they were also keen to stress the value of the unspoiled, natural sound of the voice that was free from the nervous strains of modern life. All children's voices could be heard to represent such purity, but those furthest away from the drawing room represented it best. Jean-Jacques Rousseau's views on allowing children to develop naturally were cited, interpreted, and paraphrased in various guises in manuals for vocal education, often in tandem with ideas of language development and the child's voice as a "diamond in the rough."[33]

The expertise of the old masters of canto was invoked regularly in discussions on how to best polish these natural diamonds: the methods of famous old masters like Vaccai and Porpora centered, as many of their nineteenth-century readers noted, on ideas of intuitive breathing and natural grace. At the same time, the music and performance style associated with them were understood in the nineteenth century as particularly artificial. Both the rhetoric and musical style of the seventeenth and eighteenth centuries appeared as excessively flowery and manicured to modern observers, perhaps because early modern interpretations of natural methods of singing had been expressed very differently. Inspired by natural philosophy and its attendant vocabulary, singing teachers of the seventeenth and eighteenth centuries saw the practice of song as a combination of Ars (music) and Natura (the body)—in other words, as learning to use what was already there.[34] These traditions would change considerably in the early nineteenth century, not least because the great masters of song, the castratos, lost their popularity and largely disappeared from the stage at the beginning of the century.[35] Their influence remained important, however, and their voices and methods were remembered nostalgically (or imagined grandiosely) as precious and angelic sounds of a bygone age. Musicologist and music critic François Joseph Fétis presented the loss of the castrato not as a "moral triumph for humanity," but as a "calamity for art, to be deprived of such admirable Voices" in his 1836 work aiming to make music "available to all."[36] The oratory of the eighteenth century, too, was imagined as a golden age of the craft. Lord Curzon of Kedleston, for example, in his Rede Lecture *Modern Parliamentary Eloquence*, claimed that the second half of the eighteenth century was "generally regarded as the golden age of British oratory." By the time he wrote his assessment of modern eloquence at the beginning of the twentieth century, he claimed that "men's souls are still capable of being set on fire by the spoken word; but the spark is otherwise kindled, and it lights a less radiant and consuming flame."[37]

Moreover, in addition to contemporary claims by the old masters of song and speech themselves, nineteenth-century experts often assumed that the imagined superiority of the methods of the Golden Age lay in their natural qualities—as if the old masters had an intuitive understanding of the human body, unfettered by modern neuroses and industrial artifice (but also uninformed by modern science). Rather than attributing the success of the singers and orators of yore to their knowledge, they connected it to seemingly precivilized, long-standing traditions. The artifice of early modern song and speech was artistic—and largely out of vogue in the nineteenth century—but the actual embodied practices and methods used to develop the voice were imagined to be natural by default. The old masters had unwittingly developed practices and exercises that contributed to vocal health, or so it seemed to nineteenth-century observers. The new modern scientist's task was not to change them, but rather to explain exactly how these old traditions worked on the body, how the organs were to be mobilized, and why the old practices prescribed by "the great men of old time" were so beneficial. Standing traditions in the teaching of singing and narratives about the old masters made an appearance in almost all self-consciously modern publications about vocal physiology and breath control, even though, as one author pointed out, "little, unhappily, was written down as to their methods of teaching, and practicing the art; and, in attempting to gather up the threads of their work, reliance is necessarily placed mainly on the traditions which have come down to us."[38]

The balance of trust between the authority of old masters and their venerable traditions and the new experts, with their innovative insights, would continually shift throughout the century, the tension between craft and science never completely resolved (and the methods of both also rarely completely at odds with each other). At the turn of the twentieth century for example, after at least a hundred years of growing attention to and development of vocal physiology, phoniatry, speech therapy, and other fields, Gabriel Fauré, then director of the French national conservatoire, declared that they would not be needing any courses explaining scientific approaches to music and vocality and that, as he pointed out in a 1911 letter, "the board ... has unanimously spoken against the creation of any scientific course (physiology of the voice, phonation, vocal hygiene, etc) in the conservatoire national where education needs to be and remain purely artistic."[39] His statement offers a perhaps surprising but necessary counterpoint to the story that is told in this chapter—and in this book as a whole: while the nineteenth century offers ample documentation of an increasing trust in scientific understandings of the voice and a range of embodied practices inspired by them, the perhaps equally frequent, active choice to not read these new publications, not trust science over traditions, or not change artistic or mundane practices in response to new discoveries is much more difficult to take into account. Fauré's clear rejection of the value of applying modern scientific principles to the craft of vocal music is rare, but multiple authors (and, one can assume, even more teachers of music concerned with the daily toil of making a living) simply did not engage with new scientific insights or only encountered them indirectly and in very general terms. Additionally, the more traditional belief of singing masters of the old school in the importance of practiced artifice also dovetailed with more self-consciously modern ideas about the role of experience and the skills of civilized behavior and refinement in the cultivation of the voice. Female singers and speakers in particular straddled this complex ambiguity between, on the one hand, seemingly being at a disadvantage for vocal health because of their ostensibly weaker nature and, on the other hand, appearing more at ease with speech and song due not only to their supposed innate musicality and laryngeal flexibility, but also to their cultural sensitivity. Celebrated as the guardians of culture and civilization, women were both lauded and distrusted as naturally gifted vocalists.

Gender, Identity, *Fach*

The complex question of what constituted a natural talent also influenced beliefs about the conflagration of vocal sounds and identities: despite experts' insistence on the value of education to fully develop a voice, the idea that certain bodies simply produced particular voices was widespread, even among teachers of song. In discussions of what is often called *Fach* (a vocal classification delineating different types of voices by tessitura as well as timbre), beliefs about the coagulation of identity, personality, and vocal style surfaced most obviously. In order for a certain role to be portrayed convincingly onstage, a particular sound and look were required: that the role of the young ingenue required a light soprano sound, or that grandfathers had bass voices, was commonly accepted and understood. The demands of the stage, in other words, articulated propriety largely in the same way as manuals of speech, or medical treatises, and they coincided with (middle-class) normative ideas about what the real-life counterparts of operatic personae should sound like. Ideas about which sounds were accepted as normal ones depending on the body producing them could be fluid when it came to particular roles, but also presented some hard limits, particularly when it came to gender and gendered articulations of age. Classifications of voices therefore also laid bare the ongoing tensions between nature and beauty that were at play in both vocal science and training. Throughout the nineteenth century, educators and researchers would repeatedly insist that the best speech and song would be a result of vocalizing within one's given range or timbre. Discovering what that was, however, often proved to be difficult precisely because questions of whether one should be trained as a soprano, alto, tenor, or bass were solved at the crossroads of pitch, timbre, range, and personality—in other words, the decision was a question not only of vocal expertise, but also of applying social knowledge. The roles connected to particular voice types on stage had a strong influence on what a young woman, or an authoritative man, *should* sound like; meanwhile, voice range presented practical limits to the roles available to some singers and, according to some experts, led to confusion. Clara de Chatelain, writing under the pseudonym of Leopoldine Ziska, expressed exasperation over young singers' inability to identify their own voice type in 1850, asserting: "Do not fancy your voice to be contralto merely because it happens to be limited. It is the quality and not the extent that should be consulted. I have frequently heard young ladies say 'I have a contralto, because I can only sing up to mi; but my sister has a soprano, for she reaches la.'"[40] Much responsibility for mismanaging young middle-class ladies' voices, in particular, was laid at the feet of singing masters, who were given the rather difficult task of cultivating what, in theory at least, was a product of natural aptitude and type. In addition to these seemingly fixed issues of personal timbre, similar tensions existed for the actual practice of vocalizing. To sing or speak beautifully, the logic went, one had to do so naturally, which would avoid strain and therefore also preserve the health and natural beauty of the vocal organs. This logic was hardly unique to vocalizing: in contexts like physical exercise and dance, the ideal of natural movement as a road toward both artistic beauty (or athletic excellence) and health was also frequently invoked.[41]

Although this celebration of the natural functioning of muscles and organs was often presented as a modern innovation supported by exciting new insights from physiology, phoniatry, and other empirical scientific observations, it was not a new invention of the nineteenth century by any means. In fact, it bore distinct resemblances to Jean-Jacques Rousseau's educational ideals, according to which children—including their voices—were to develop naturally. Rousseau's work was cited regularly with regard to vocal health and especially the threats of overcivilization. French physician Félix Voisin, for example, called upon Rousseau's ideas to underline the value of a simple, rural life as a safeguard against stammering, thus calling forth an image of a "noble savage" whose vocal practice was beautiful and healthy precisely because of the absence of artistic or educational intervention.[42] The fascination of both medical and musical experts with the vocal purity of the noble savage also explains, perhaps, the particular obsession many authors seem to have had with boys' voices, which—more than the sounds of young girls—were understood as especially precious because of their short-lived existence. The attraction of boys' voices, too, was nothing new in the nineteenth century, and would last far beyond it.[43] The fragility and short-lived nature of the treble voice continues to exert influence over audiences today, much as it did in the time of Bach's choir, and yet very differently too. The very particular image of the modern treble choir owes much, I would argue, to nineteenth-century explanations for what distinguished male and female voices, or rather male and female larynxes, from each other. Most experts stressed that there was no anatomical difference between boys' and girls' voices in childhood, but there nevertheless seems to have been a widespread belief that their voices prefigured the adult sounds they would eventually project—as if their young bodies had already molded themselves into the shapes that would be socially and culturally required of them. The sound of the feminine adult voice, heard as fuller and more rounded, was thought to be prefigured in girls' voices much in the way girls' bodies prefigured the adult, potentially maternal, ones they would come to inhabit. According to French physiologist Édouard Fournié, this was due to the influence of girls' reproductive organs on both their bodies and their moods from a young age: "Driven by her new destiny, the young girl already seems to foresee the important role that the uterus will play in her physiological and pathological existence. Sad, dreamy, with a languorous and veiled gaze, she seeks solitude to explore new and unknown inner sensations."[44] The move toward adult gendered identities was experienced and represented as a transformation, including important vocal changes, as children's voices were understood to be mostly identical in both genders. "Until the age of 13 or 14, the young girl's voice differs little from that of the young boy. Its timbre and range are more or less the same in both sexes. But from that moment onward, differences will arise. The young boy will become a man in his voice, and the young girl will become a woman in hers. That supreme moment is the age of puberty."[45] The process of change that occurred in youth—changing girls into mothers and boys into men—was of special interest to vocal experts, since it could be such an eminently audible one. Especially in boys, the voice break presented a clear moment of maturation—a mutation from one identity into another. Physician, "anthrophonist," and amateur singer Carl Ludwig Merkel noted, for example, that he had gone from being a soprano to being a baritone.[46] This change, particularly in men, signaled very publicly that a temporary state of boyhood had come to an end. It also started to lead to ideas, like the one expressed by Fournié above, that there was a particular stage in life in which such changes took place—a stage that was understood as biological, to be compared to the development of all sorts of animals, but that was also cultural and social (and therefore different) for humans. "Several animals use their voice only during the season of love; the birds which sing all the time have, during that period, a more powerful and sonorous voice. At the time of puberty a great change takes place in the nature of the sounds produced by the vocal organs."[47] In boys, this was assumed to be a particularly difficult transformation. Whereas girls' larynxes grew only slightly larger to turn into women's, the changes wrought on boys' voices were seen as much more significant, arousing anxieties about their maturation. Changes in expectations surrounding the sound of the adult male voice point in a similar direction: they show how vocal production can be read as an expression of a changing gendered identity. The moment of the voice break had long been established as a moment of significant change, but ideas about what should lie at the end of that transformation changed over time. Likewise, the nineteenth century's changing expectations surrounding masculinity led to changes in its sounding realities—some more obvious than others. It would not be until the end of the nineteenth century that the Wagnerian tenor would come to symbolize the kind of physically

Dimensions du larynx chez l'homme[1].

AGE.	DIAMÈTRE vertical.	DIAMÈTRE transversal.	DIAMÈTRE antéro-postérieur.	Grande circonférence.
	m.	m.	m.	m.
27 ans	0,045	0,042	0,038	0,142
30 —	0,048	0,048	0,035	0,143
38 —	0,042	0,051	0,033	0,140
42 —	0,042	0,040	0,035	0,130
45 —	0,045	0,040	0,036	0,136
50 —	0,043	0,044	0,039	0,134
56 —	0,043	0,040	0,040	0,133
60 —	0,045	0,043	0,034	0,131
Dimensions moyennes....	0,044	0,043	0,036	0,136

Dimensions du larynx chez la femme.

AGE.	DIAMÈTRE vertical.	DIAMÈTRE transversal.	DIAMÈTRE antéro-postérieur.	Grande circonférence.
	m.	m.	m.	m.
24 ans	0,036	0,042	0,025	0,115
25 —	0,035	0,040	0,024	0,107
30 —	0,037	0,042	0,027	0,117
34 —	0,040	0,039	0,026	0,108
38 —	0,035	0,044	0,024	0,109
40 —	0,040	0,046	0,027	0,128
50 —	0,034	0,041	0,028	0,106
70 —	0,035	0,037	0,026	0,108
Dimensions moyennes....	0,036	0,041	0,026	0,112

[1] *Traité d'anatomie de Sappey*, t. III, p. 369.

Figure 5. Édouard Fournié, *Physiologie de la voix et de la parole*, Paris: Delahaye, 1866, p. 150. Courtesy of the Wellcome Collection.

strong and heroic masculinity that would become a hegemonic model until it ran into its dramatic, culminative frustration in 1914. However, prefiguration of the heroic tenor arose earlier in the nineteenth century. Not only were male altos and castratos disappearing from the stage,[48] the light haute contre was going the same way, and ideals of strong belted tones started to replace them.[49] The *ut de poitrine*, a strong belted high C, became the defining feature of this type of masculine voice in the latter half of the nineteenth century,[50] but a focus on vocal strength in men was apparent in physiological treatises on the voice in the first half of the century.

It is impossible to separate the emerging therapeutic thinking about puberty as a distinct period in life, and in the development of children's voices, from the role vocal health, propriety, and beauty played in sociocultural constructions of identity. Voices, like other corporeal characteristics, are made, not born, as numerous historians and musicologists have pointed out.[51] If a voice was really only made after puberty, as nineteenth-century experts seem to have taken as writ, vocal identity necessarily appeared not only as an intrinsically adult affair, but also as one inflected by gender and class (and, as will be addressed in Chapter 4, ethnicity). The made voice, whether fully trained or molded merely by the daily exercise of mundane practices of speech, therefore existed at the intersection of what appeared as immovable biological facts and conscious individual practice. Moreover, the voice itself had a fluid character: it changed over time and had the ability to shape-shift at its owner's will (or at least with intention and practice). As a result, the articulation of complicated individual and collective identities through voice was mired in anxieties about authenticity and health.

Although these anxieties centered around pathologies and improprieties of the adult voice, children were targeted particularly in attempts to cure any vocal anomalies. Imagined as more pliable—both symbolically and physically—children's voice boxes became a prime site of tension around ideas of purity and health, and their intrinsic connection. It was therefore that song needed to be taught at "that poetic age at which the heart is formed," so that good habits could take hold of both their bodies and their hearts.[52] Aging, as a process, was therefore imagined as both one of development and one of decay—either way, the aging process constituted a process of ossification. In current studies about the gendered and ethnic associations attached to voices, the hardening of expectations regarding binary notions of identity as expressed through the voice has been understood as ossification: a fixing of fluid cultural categories in human throats, demanding that particular people sound a particular way.[53] For the nineteenth-century larynx, however, ossification was not thought of as a cultural issue, but a physical one.[54] Adult male voices were imagined to be more rigid—the cartilages harder, the shapes less flexible. Together with its greater size, this made the adult male larynx appear as the more developed one, an image that neatly dovetailed with expectations of masculinity in other arenas of life: more than a stiff upper lip, the stiff, unpliable larynx articulated ideals of masculine self-control and steadfastness.[55]

The ossification of the voice, in other words, had distinctly different effects on the social status of the speaker or singer and the general perception of them as dependent on gender. In fact, aging was expected to sound very differently in men and women in general, demanding pliability in women as long as possible—the loss of flexibility in their voices late in life was read as a sign of a loss of femininity. Or, as Colombat put it, the female voice "almost always preserves the characteristics of infancy," and a loss of youthfulness in women was heard as anomalous.[56] Elderly male voices, by contrast, would lose the steady footing of adulthood late in life, and go back to childish instability.[57] More than to any clear biological change, this development, like puberty, seemed closely related to the social function of the gendered aging voice: the sound of stability, symbolized by a low pitch, a steady rhythm, and modest and controlled modulation, signaled a voice's suitability for public speech. The unsteady and changeable sound of feminine and infantile throats, on the other hand, was designed for domestic use.[58]

Although these operatic ideals may have made little difference to most men's daily lives, the proto-eugenic ideas that underpinned them filtered through to medical and therapeutic writings far off the stage, where men were exhorted to develop strong voices. Descriptions of the vocal organs, in particular, translated ideas about masculine strength and control into anatomical features that would define vocal sounds. In Colombat's introduction to the larynx, expectations about male and female bodies were thoroughly intermingled with cultural norms about masculinity and femininity: "The larynx is much more developed and prominent in man than in woman, in whom this organ has but two thirds and even half of the volume of that of man. In the latter, the retreating angle of the thyroid cartilage is acute, while it is rounded in woman, in whom the central slope of the superior border of the same cartilage is less deep, and the epiglottis less large, less thick and less prominent than in man."[59] Male larynxes, like masculine voices and masculine behavior, were represented as more defined, more prominent, and more robust than women's. The ideal was kept up through the discipline of women's seemingly more rounded voices as well, which were understood as constituting the other side of a strictly defined binary—or, as *Hamilton's Modern Instructions in Singing* put it rather more succinctly: "The voices of adults, according to their sex, are divided into two classes, male and female."[60] Children, and the exceptional voices of castratos, complicated the image somewhat, but did not call into question this dichotomy: "Acute voices belong to women, children, and eunuchs; grave voices, to full grown men."[61]

Various visualizations of male and female tessitura depicted the female voice as clearly distinct from the male voice as well. The binary was thought of as an almost mathematical certainty, with female voices imagined as naturally pitched an octave above male ones.[62] "It will suffice to note," Stephen de la Madelaine concluded in his rather lengthy explorations on gendered pitch and range, "that men sing one octave below women."[63]

This octave of difference supported a binary that signaled not only different biological roles, but different social ones as well. Research on operatic music has pointed to ways in which the disappearance of the castrato and concomitant rise of the tenor onstage served to strengthen ideas about heterosexual couplings as a natural affair growing from the distinctly different bodily features and cultural roles of men and women. As these therapeutic and scientific documents show, this vocal logic went far beyond the stage, and beyond acoustic expression: the difference (in pitch, but also timbre) between male and female voices, in the nineteenth century, came to be understood not only as a symbol, but as a constitutive part of gendered identities. And like other aspects of those identities—like physical strength or maternal instinct—the gendered voice was presumed to be both preexisting or naturally occurring in the body, *and* in need of constant practice, development, and preservation. Many manuals for vocal training, like James Mackness's work on hoarseness, claimed not to "treat of the methods by which the public speaker may best cultivate his powers, but to point out how he may best preserve them from disease or regain them when impaired."[64] In other words, to support a socially and artistically proper voice that was presumed to be naturally present in the body.

Materiality and Metal

The ongoing tension between images of the voice as given and naturally embodied or as laboriously acquired and performed is still a familiar one today—narratives of talent and tales of excruciatingly hard work suffuse the worlds of both song and rhetoric. The social implications of that tension also persist, even if the gender roles articulated onstage and in life have changed over time. The modern imaginary of the voice as a physical, tangible thing (as described in Chapter 1) plays an important part in how this complex edifice of vocal propriety is given shape in practice. Whether we speak of the orator using his "silver tongue" as a tool to convince or move his audience, or envisage the singer's vocal organs as a precious musical instrument in need of careful handling, the focus on the material and physical nature of the human voice may hide the socially enacted character of vocality, but in fact coproduces the social categories within which vocal propriety is construed. Moreover, speaking about how different vocalizers "use" their voices differently enacts a separation between the speaker or singer and their embodied voice. This voice, as a sounding and material something, travels to the ears of a listening audience and has an impact on others; but the body of the vocalizer—while not unaffected either—stays securely where it is.[65] Moreover, as a thinking, volitional being, the singer or speaker is supposed to control and manage the voice. Playing it, in other words, as one would play a musical instrument.

In the remainder of this chapter, I want to explore how this imagery of the voice-as-instrument was mobilized by vocal experts in the nineteenth century to articulate ideas of civilization and social difference through vocal sound. Musicians and natural philosophers had long contended that musical instruments, in their various guises, all presented imperfect attempts to imitate that greatest of all musical devices, created by god(s):[66] the human voice. This discourse resurfaced in the nineteenth century, but was rephrased to suit the vocabulary of newly developed ideas about acoustics and physiology. In his handbook on elocution and reading aloud, author and dramaturg Édouard Mennechet stressed the divine origins of the human voice: "All the musical instruments that charm our ears have been created to resemble the vocal organs, but the vocal organs only resemble themselves, they are the work of God. If man cannot even understand them, how should he imitate them?"[67] At the same time, the culture of objectivity that arose around studies of the human voice gave rise to the mirror argument: discussions on the physiology and physics of vocal sound often clustered around an understanding that vocal mechanics were to be analyzed along the lines of artificial, instrumental sound production. If similarities between the workings of the larynx and the trumpet or between the vocal cords and the strings of a violin were discussed, this occurred no longer in debates about instrumental imitations of nature, but rather as a way to support understandings of the inner workings of the voice within the body. Singers and speakers were not only expected to care for their voice as if it was a precious object and to play it like an instrument, they were also taught how to know their voice as they would an external entity: as tangible and open to objective observation.

Moreover, this likening of the voice to musical instruments went far beyond ideas of symbolism or metaphor. Musical instruments were used as comparisons to the vocal organs largely to allow authors of manuals (and their readers) to more fully understand the complicated workings of the voice—not just its mechanics, but also its physics and acoustic properties. Exactly what those mechanics and acoustics were, however, was a matter of debate.[68] Therefore, questions of which instrument was most suited to explain the workings of the voice remained a source of discussion and confusion for most of the earlier nineteenth century. The main issue in these questions was whether the voice was to be understood as a string instrument or a wind instrument. In their 1885 *Physiologie de la voix et du chant: hygiène du chanteur*, Achille Gouguenheim and Marcel Lermoyez painted a vivid account of historical comparisons between vocal physiology and the mechanics of various instruments, starting with Galen, who, apparently, "finds that the larynx is surprisingly similar to the flute,"[69] then moving through the eighteenth century with theories comparing the workings of the voice to the "strings of the viola," and the early nineteenth-century notions that saw "the assimilation of the larynx with brass instruments." Gouguenheim and Lermoyez eventually settle on the pipe organ as a fair representation of the human voice.[70] This may seem like a somewhat frivolous debate to carry so much weight, but it does in fact shine some light on how the vocal organs were defined and delineated by different authors and in different contexts: comparing the voice to a string instrument meant centering the role of the vocal cords in sound production, for instance. The comparison reflected, perhaps, the almost obsessive interest in these frustratingly hidden and seemingly invisible organs that scientists were so eager to lay bare in the early nineteenth century, but also came under scrutiny, as the mechanical imagery of the so-called voice box seemed to be too limited to explain vocal sound. Thinking of the voice as a wind instrument opened up a much more holistic understanding, in which the lungs and diaphragm were included as vocal organs.

The difference of opinion (or imagination) regarding strings versus wind would, for some authors at least, eventually be resolved by looking at other instruments. Marc Colombat de l'Isère concluded that the trombone was the most suited to embody the characteristics of the human voice.[71] Teacher and professor of eloquence Julien Duquesnois would settle on the image of the clarinet, a woodwind that combined an obvious vibrating element (the reed), which gestured toward the function of the vocal cords, with an obvious place for the lungs and displacement of air.[72] While both those images seem to have been introduced as compromises between the "voice as violin" and "voice as trumpet," they also seem to indicate a move away from a strict distinction between anatomical and performative or therapeutic images of the human voice, and an attempt to resolve the tensions between "natural" aptitude and training. Combining an insistence on the whole human body as a sounding, vocal instrument with a focus on the need for physical and continued work to make sound, these instrumental compromises developed in step with the rise of new science-based but educational models for the development of the naturally given human voice. There is, however, no simple chronological narrative to be told about these comparative imageries of musical instruments.[73] Nor can simple conclusions be drawn about cultural distinctions or contexts, although the image of the voice as a musical instrument seems to have been somewhat more prevalent in French publications than elsewhere. This may have been connected to a French commitment to artistry and cultivation over scientific insights on physiology in institutions like the Conservatoire Nationale. The strong association of the conservatoire with its national task—that is, its role to cultivate and maintain a French musical culture—was part of a wider interest in language and voice as a French/national question,[74] and seems to have gone hand in hand with both a certain distrust of exclusively acoustic inquiry, and a stronger reliance on *anthropologie* as the emerging discipline in which discoveries about language, speech, and voice were situated.

By focusing on the connection between the art of music and the workings of the voice, publications that centered narratives of vocal physiology framed analogously as musical instruments also contributed to the development of discourses surrounding skill and virtuosity. If the voice were to be played like an instrument, the qualities of the musician in charge were of equal importance to the qualities of the instrument itself. These categories were, for singers and speakers, even more blurred than for instrumentalists, as their voice *was* effectively their instrument, but seemed to carry largely similar connotations. As James Davies has pointed out in *Romantic Anatomies of Performance*, the embodied practices of highly skilled (and very successful) musicians were the result of careful training, inspired by both scientific theories and various, almost magical notions of talent, soul, and personality.[75] Extremely virtuosic or powerful performances by a recognized diva in particular became a special obsession in the nineteenth century.[76] These images of exceptional singers who enraptured their audiences in hypnosis-like trances[77] paint a picture of the ideal singer as a force of nature, even though it seems to have been clear to all that that was a carefully cultivated fiction.

The precarious balance between talent and practice was supported in this new imagery of the vocal organs precisely by the complicated relation between the mechanism of the larynx and its context in a living, breathing, and performing body. In fact, the context in which nineteenth-century experts placed the human larynx was often even more broad and more complicated than that: the production of human vocal sound was understood in the context of acoustics.

Rather than seeing the voice as a mechanism, then, these authors placed the lungs, vocal cords, and vibrations in the realm of physics—focusing on minute changes and barely perceptible movements rather than on the rather more clunky actions of muscles and lungs for their own sake. The machinery discussed in the previous chapter helped in this endeavor by making not only the vocal cords themselves but also their otherwise invisible effects available as an image to analyze.

This imagery was far from innocent, and not as objective as its producers liked to believe. The visual language produced in those manuals and treatises also served to communicate a repertoire of aesthetic expectations that was believed to express social distinctions (like distinguishing features connected to gender, age, or social status, as demonstrated earlier in this chapter). The instrumental imaginarium produced in the first half of the nineteenth century could serve a similar purpose—albeit often in less obvious or binary terms. Likening voices to a range of musical instruments, like visualizations of the throat, gave an impression of distance and neutrality, but also articulated a range of unspoken expectations. In conjunction with the imagery of wind, brass, string, and other instruments, many authors also had ideas about timbre that they expressed with material metaphors. Playwright and critic Ernest Legouvé, who was cited in numerous works on the voice, noted that anyone who lacked "metal" in the voice "would never shine as a colourist." The metaphor of color had been applied to sound and music before (the German term for "timbre" explicitly posits *Klangfarbe* as a matter of color). In the nineteenth century, this metaphor would increasingly become the topic of more specialized discussions on both poetry and synesthesia. Tapping into this ever more concrete belief that sounds—both musical and linguistic—had "color,"[78] authors used material qualities as a way to describe timbre. According to Legouvé, "The metal may be gold, silver or brass; each has its individual characteristic. A golden voice is the most brilliant; a silvery voice has the most charm; a brassy voice the most power. But one of the three characteristics is essential. A voice without metallic ring is like teeth without enamel; they may be sound and healthy, but they are not brilliant."[79] Although the narratives around voices like instruments were not explicitly linked to these theories of metal in the voice, it is likely that readers took such ideas for granted and applied them to their reading of the physiological and acoustic chapters in the same manuals. In a more indirect sense, such implications could also translate into identity-based expectations for different voices. The brassy voice, for example, would gain a different meaning depending on whether it was launched from a male or female body—and would eventually come to be connected to the mostly ethnicized voices of jazz singing.[80] Voices without metal, according to Legouvé, carried an undefined softness that hinted at a lack of definition in the singer's personality as well. "Last of all is the velvety voice. This is worthless if not allied with one of the three others. In order that a velvety voice may possess value, it must be reinforced (doublée) with 'metal.' A velvety voice is merely one of cotton."[81] There was no question that the velvety voice was not only deficient in aesthetic quality, but was also heard as socially improper. Its lack of clear material definition or relief equated to a refusal to express any social or cultural distinction: by not giving shape to a clear role or character, it also failed to signify an identity. What these voices without "metal" show, too, is that despite the rising focus on physicality and the materiality of voice in the nineteenth century, voices never became "just" body or health: their physical status was codetermined by their social salience. In other words, to be "sound and healthy" outside of cultural and social distinction was simply meaningless.

INTERMEZZO 2

A Stern Voice from Abroad

Corresponding with Nina d'Aubigny

September 15, 1807. The *Glory*, under the command of Horatio Beevor, has set sail for India. It will arrive in the harbor of Kolkata in March 1808, delivering new English blood to the colony. On board, reluctantly sharing the cramped space of the ship with her British travel companions, we also find Jana Wynandina Gertrut d'Aubigny von Engelbrunner, or "Nina," as she liked to be called. If her diary is to be believed, she must have been a rather unusual passenger. Before sailing off she had consciously purchased a couple of bland, dark dresses so as to not give the impression that she was traveling to the Indies in search of a husband. Because she was thirty-three and unmarried, with a long history as a governess in various aristocratic households, it is easy to see how she would have been mistaken for one of the many not-so-young girls carted off to the colonies to narrowly escape spinsterhood. That is, at least, how she described her fellow female travelers on the *Glory*:

> This is their general belief: a girl would only go to India to find a husband. This idea has taken root so firmly, that it would be pointless to hope they would accept any other reason. And since it is also customary amongst the English to send their superfluous and ageing daughters and sisters far away to the market, it is usually the case that these women cannot start their plans of conquest soon enough and, as soon as they arrive on the ship, they aim their batteries to enter into battle.[1]

Nina had other plans, however. She rather fancied herself to be a cultural connoisseur and had artistic and anthropological interests—rather than marital ones—to pursue in India. Having already published a number of travel accounts of less exotic voyages (to the Netherlands and England), she must have imagined that she was more experienced and more cultured than a lot of the English colonials she met in Kolkata, whose insights on the natural and cultural splendors of India she found deeply disappointing: "Now one would think it would be easy to find good conversation, as all these things about Asia are bound to be interesting. Only, I have not found anyone who has anything more to tell than stories about sedan chairs and tiger anecdotes."[2] She was also less than impressed by the way her fellow British colonials (and presumably her hosts in India) treated the local population.

It is easy to imagine Miss d'Aubigny with a perpetually raised eyebrow, or with a slightly disapproving smile . . . an advantage perhaps for a single woman making a living raising and educating various aristocratic girls. And yet, she wasn't just a governess. As a young teenager, she had received musical instruction from Italian composer Pompeo Sales and was trained as a classical singer. She performed with her sister to great acclaim, but to fairly small audiences, as the Aubigny sisters were not supposed to become professionals of the stage. The cultivated voices just came with their identities as young ladies of leisure. Nina did seem to have been particularly gifted, and managed to supplement her income in India, when necessary, by returning to the stage. More importantly, she made a point of providing her young charges with a musical education as well, and would eventually make a name for herself (if only posthumously) as the author of a valuable manual for the musical education of children in general and girls in particular within the household. Insisting that girls' education in music was not only relevant but even necessary, she casually dismissed the notion that the role of housewife was to be restricted to its practical components, and connected the image of the well-educated woman to the notion of the civilized nation.[3] Although a woman's fate in Germany was a reasonably lucky one, she argued, education in the arts in general, and music in particular, was in need of improvement: "But even though we have made great strides [in women's education and the development of music as an art] an egregious lack remains concerning the cultivation of the organs."[4] According to her biographer, Manfred Ensberger, the great Ludwig von Beethoven must have approved of some of her ideas—or at least must have been aware of them: his library contained a copy of d'Aubigny's *Briefe an Natalie.*

The full title of the work was *Letters to Natalie About Song, as a Way to Further Domestic Happiness and Communal Joy. A Manual for the Friends of Song Who Want to Educate Themselves, or for Mothers and Educators Who Want to*

Build Their Charges in This Art, betraying its domestic and educational goals.[5] The *Briefe* of the title point to the genre of the work: d'Aubigny did not write her musical instructions as a manual, but as a collection of letters to an imaginary female friend. Informative, romantic, or moralizing fictive correspondence was a popular genre at the time:[6] it was used in autobiographical, educational, and travel writing, and the *Briefe an Natalie* belongs in all three of these categories. The letters reflect on their author's own life and experiences, betray her internationally mobile past with references to national specificity in the art scene and international comparison, and are explicitly instructive in tone. Much like in other memoirs of the time, the author claims that she has written her work at the request of others, and in response to an existing need for her accumulated knowledge. And even though the *Briefe* were written at a fairly tender age and are not explicitly addressed to an audience of young readers eager for the wisdom the wise author can convey, the conversational tone directed at the imaginary Natalie has an almost maternal ring to it, reminiscent of memoir writing. From the very first letter, the goal of this pseudo-correspondence (Natalie never writes back) is set out. D'Aubigny wants to make the virtuoso education that had been on offer to male singers since Tosi and Agricola available to women, a sex "not lacking any love of song, but rather systematic instruction."[7]

D'Aubigny sets down a highly sophisticated education, one more commensurate with her own upper-class notions of upbringing than the kind of patriotic musical education that would later become part of a (proto) national endeavor toward *Volkserziehung* (popular education).[8] Nevertheless, a number of the beneficial effects that are forwarded by d'Aubigny—and that presumably legitimized her claim for the necessity of musical and vocal education—ring familiar to those at home with late nineteenth-century pamphlets on (German) health and the practice of singing, or indeed the school manuals of the same period. In the *Briefe an Natalie*, d'Aubigny addresses what would become a common thread in the work of music educators and physicians alike. She lays out the tensions between the idea of a stable, all-defining nature (biology) in which perfectly crafted organs consistently produce a natural and therefore perfect voice on the one hand, and the need for constant cultivation, resulting in anxious practices of exercise and the avoidance of pathology, on the other hand. In d'Aubigny's writing, the practice of singing is presented not only as the result of physical perfection, but also as its origin. Singing would contribute to the development of children's organs and therefore play a part in the care for their physical and moral health. It was to be taught as soon as possible, ideally even "in the cradle."[9] Simultaneously, she claimed that young children were natural singers; the antithesis of good vocal education was therefore not a rough or uneducated natural state, but rather the cultivation of "unnatural" tendencies to lose one's voice.

> It is really a case of neglecting the most beautiful gift, when one hears so often: I do not have a voice. Has nature not kindly gifted every being with a voice to speak? Does not every free child sing in the street, when cheerful? And when was then this certain period when, after the laws of this kind nature, so many people had to lose their voices, never to regain them again, as one could be tempted to conclude from the rare use of the voice in song?
>
> If one would keep a child from ever using its feet for the first years of its life, it would probably say just as convincingly, I have no feet! This is quietly admitted to be the case for many fashionable ladies, whose feet have been replaced by their horses, and the Turks on their pillows prove it to be true as well, when they see Europeans walking around without a special destination and think them ill or mad for it.[10]

She makes the case in a distinctly feminine fashion, for an explicitly feminine world. The letters are, of course, addressed to the Natalie of the title, and d'Aubigny repeatedly makes reference to the fact that this is a correspondence between women, addressing Natalie as a dear *Freundin.*[11] Moreover, the book is dedicated to a number of well-known, titled women, positioning it squarely in the sophisticated, semipublic sphere of aristocratic domesticity (and possibly setting it up as an aspirational document for women in the wealthy middle classes hoping to raise their and their daughters' profiles through artistic cultivation). The title locates the proposed education firmly in the home, and in the hands of governesses and mothers, but this centrality of motherly care is less present throughout the text of the book, in which mothers seem to play a very minor role indeed. D'Aubigny is not advocating for an education in nursery rhymes and lullabies, or any constellation in which traditional notions of motherly care, nursing, and music would converge, but rather for an education in song that would be "influential for the physical and moral education of the female sex."[12] Moreover, her choice of music is rather more ambitious, and the goal of musical education is not an emotional but an artistic one. At the heart of her argument lies the development of good taste, of impeccably performed music composed by accomplished artists and appreciated by an educated, critical audience.

Although the author presents herself mainly as a woman of good taste, and denies being interested in the direct usefulness of the education she proposes, her claims can be read as a discourse directly preceding the professionalization of vocal experts in the nineteenth century. Like her male colleagues, d'Aubigny stresses her own place in a long tradition of musicians and singers—sprinkling her letters with accounts of her meetings and lessons with the great names of her own time as well, thus showing off her place in an existing network of artists. Her method is not a scientific one (and does not pretend to be so), but it is based on extensive knowledge and erudition exhibited through a number of considered observations. In her *Briefe an Natalie*, she seems to foreshadow what the fields of ambitious amateur singing and vocal care would later become: the terrain of highly specialized women.[13] Her work lacks the scientific and empirical rigor of what Emma Seiler would provide in *The Voice in Singing* and the explicitly therapeutic goals of, for example, Kate Behnke, but despite its conversational tone it does not lack systematic thought—or indeed professional confidence.[14] D'Aubigny writes of the great musicians she meets with admiration, but without an inclination to submission, clearly setting out her own educational agenda.

Part of d'Aubigny's claims to professionalism—and part of what made her so convincing as an expert—is her broad international perspective on the art world. She attributes these insights mainly, it seems, to her own aristocratic background. Her family hailed from Kassel, but regularly traveled in the German states and the Netherlands. In 1794, she followed her newly married sister and her young family to Bückeburg, and later she would continue to travel (whether out of a sense of adventure or out of financial need is not entirely clear). Aside from her own international mobility, d'Aubigny seems to have inhabited a somewhat cosmopolitan world, peopled by guests and visitors from all over Europe and filled with cultural knowledge and artifacts far beyond the confines of what would later become Germany. In her letters, she explicitly referred to musical traditions she defined as somehow foreign, but still related (and therefore valuable) to the German context. The Italian style of singing appeared, as in most European work in the nineteenth century, as the oldest and most natural one, but d'Aubigny hastened to point out that it was not necessarily superior to German song—or, rather, that German music would have the potential to supersede Italian song if only the musicality of the language would be well attended to. The "recognized riches of the German language" would hardly hamper vocal education, she surmised, if only poets would work to "take song into consideration" when writing.[15] Her appreciation of a German art is at best a proto-national one, however. D'Aubigny inhabited a world consisting of a variety of regions, cultures, and dialects that had not yet been organized into neatly divided nations, even if she occasionally uses the term to describe a people sharing a language. Whereas political radicals of her time (such as music educator Joseph Mainzer) saw the children of France and Germany as the future of a politically defined people, d'Aubigny saw distinct musical traditions connected, but not confined, to culturally differentiated regions.[16]

Her well-traveled, aristocratic personality might have played a role in that perception of the European landscape as well. Although her scathing criticism of the English (and her patronizing views on the Indians) had a distinctly ethnic taste, she seems to have based most of the distinctions she made between the peoples of Europe on their language, both when traveling and in discussing their musical cultures. Writing at a time when the standardization of languages was taking flight, her use of the term "German" is particularly interesting.[17] Whereas in her travel writing she shows herself to be keenly aware that spoken language has variations that are not accounted for in the written text (and in foreign language education), she does seem to assume that in German, only one type of pronunciation could be considered correct. She does not specify what her concrete ideas on pronunciation were (presumably, she assumed her own pronunciation of the language was inherently correct), but a number of theatrical and musical manuals would appear throughout the nineteenth century to help singers, actors, and the intended audience of the *Briefe an Natalie* to eradicate local, dialectic sounds and replace them with proper "high" German.[18]

Despite the absence of obviously political notions on language and nationalism (as expressed by writers like Mainzer), and despite her general stance toward a cosmopolitan, international world of music and art, d'Aubigny's writings point toward the connections between biology, nationalism, and song that would soon develop and become iconic for modern singing. Throughout the nineteenth century, the particular tension between a voice lodged in a naturally given organ and the culturally defined practices of cultivation would lead to investigations far more specific and far-reaching than that of our well-read and experienced domestic singer. D'Aubigny's interpretation of vocalized Germanness, deeply attached to her mother tongue but in constant dialogue with the linguistic and musical traditions of Italy and France, would make way for more hard-lined dichotomous (and often ethnic) interpretations of national difference. In 1904, another multilingual and mobile German amateur singer, Emile Sutro, would spell out his somewhat eccentric views on the matter in *Duality of Thought and Language* as follows: "I speak English, hence I am an Englishman or an American—in fact, an Anglo-Saxon. I belong to the Anglo-Saxon race. Or

I speak German, hence I live and think like a German, etc. I belong to the Germanic race."[19] And although speech, for Sutro, was a matter of language as well as voice, a combination of neurological and anatomical elements, it was above all connected to moral and emotional characteristics that were ingrained and geographically specific (even after migration).

Sutro's brand of internationalism and his notions on the neurological grounding of linguistic and vocal differences would have been completely foreign to d'Aubigny. And yet, Sutro grapples with the same problems of natural physicality and cultivation of the voice, and with the tension between the universally human capacity to sing and the specifically German ways of doing so. In a way, they respectively mark the beginning and the end of a period of almost unique certainty about the materiality and humanity of the voice. Both before and after d'Aubigny's and Sutro's writings, the voice was heard as distinctly eerie and potentially disembodied.

CHAPTER 3

Doing Voice

Expertise, Experience, and Expression

> Let the reader do as follows: (1) Place the finger on the shield cartilage, and press it vigorously backward (2) Sing loudly any high tone that is well within your compass. Hold this tone steadily, and be quite sure you do not alter its pitch. (3) Now suddenly remove your finger, continuing to sing as before. What is the result? Your tone is raised by a third, or even more, according to the amount of pressure you exercised on the shield.
>
> —Emil Behnke

In 1882, vocal therapist Emil Behnke encouraged his readers, with the help of a number of anatomical plates, to test his vocal theories for themselves. Observing visual renditions of the voice box, and maneuvering their fingers around their own throats, they would be able to fully understand the effect of the manipulation of different muscles, cords, and ligaments on the sound of their voices. Behnke's manual was new, but greatly relied on his readers' existing ability (or willingness, at least) to picture the inner workings of their own throats post-1855.

Garcia's hotly discussed inventions in the middle of the century may have been heralded as pioneering and groundbreaking work, but by the time the laryngoscope came into use, the field of laryngology was more or less established. Around the 1860s, the throat had been medicalized and its anatomical features charted. Although the territorial disputes between physicians and elocutionists would not disappear anytime soon, they certainly seemed to become more nuanced. More importantly, perhaps, experts from both sides of the field were increasingly working together—advising teachers, patients, and singers, but also co-writing manuals. Emil Behnke, for example, who presented himself as a "teacher of voice production," co-wrote an often reprinted "practical guide for singers and speakers" with aural surgeon Lennox Browne. In their book, the results of scientific experiments, practical breathing exercises, photographs of larynxes, and dietary hints were intertwined to produce the "combined view of vocal surgeon and voice trainer."[1] Their work was clearly aimed at laymen, exemplifying a more general "scientization" of society (or at least of the professional classes).[2] However, the cooperation of specialists from different disciplines went beyond a mere trickling down of scientific novelties into the broader discourse of general manuals. In the introduction to their "practical guide," Behnke and Browne launched a plea for vocal physiology, asserting that anatomical medical knowledge was indispensable to the artist. Their argumentation, however, was not based on the authority of science per se, but on that of the history of canto: the Old Italian Masters were indicated as torchbearers for the study of the "physical laws" that were thought to govern the singing voice.[3]

Behnke and Browne were hardly the only ones to appeal to the authority of art and tradition. Singers and teachers of canto had a long tradition of claiming an affiliation to their preceding masters or adherence to specific schools of singing—sometimes going as far back as greats such as Porpora or Lamperti.[4] The theme of fathers and sons, either metaphorical or real, runs through the field of vocal knowledge like a red thread. While Garcia Junior and Senior may have been part of a long and illustrious tradition of singing dynasties, other pairs were very much part of the brave new world of science and therapy, where individuality would be expected to take precedence over family connections. The Bells, Benjamin Beasly and his son-in-law, Dax père and fils, Isaac and Lennox Browne . . . the collaboration and practices of intergenerational transfer were a feature of all levels and corners of the field, from respected (and degreed) medical professionals to downright quacks. The relations between these pairs were not necessarily smooth, or always characterized by benevolent mentorship, however. Lennox Browne, for example, seems to have been pushed into the new, less prestigious field of laryngology because of his father's tarnished reputation in the medical world. Mentorship was also not entirely exclusive to the patrilineal line: the Behnke family represents an important part of a new school of thinking on vocal education and elocution that seems to be at the base of what is now the heavily female-dominated field of speech therapy. Nevertheless, the image of benevolent fatherhood and the idea of a legacy being passed on to the next generation played an important role in the imagination of the fields of vocal knowledge as they became more established. Inventions like the laryngoscope or discoveries like the speech center seemed to have many (real and metaphorical) fathers who jostled for primacy (see Chapter 2), and the scientific and therapeutic practices surrounding them boasted a number of godfathers in addition to those contenders.

The centrality of "fathers" of invention is, perhaps, not particularly surprising in a world in which expertise was coded as masculine[5] and the leap from begetting offspring to begetting knowledge and ideas seemed eminently logical. However, the stress on fatherhood also points toward the particular intersectional understandings of middle-class masculinity in the nineteenth century. That is, it was always intrinsically bound up with maturity—or, to put it differently, masculinity was defined in opposition not only to femininity, but also to childishness. In the construction of expertise in general, and vocal expertise in particular, this function of age/maturity as a necessary qualifier of masculinity was central, because the voice was understood as nature in need of cultivation—and cultivation took experience, and time. This was reflected in observations and analyses of the voice itself as well as its producing organs. Borrowing terminology from the nascent field of evolutionary biology, male larynxes were imagined as more developed than those of both women and children, conferring unto men a sense of vocal mastery that spoke of more intergenerational skill, and of access to a larger cultural legacy. Imagined as fathers, male speakers and singers took their place in a long line of skillful vocalizers. If the self-help manual and the can-do spirit of laryngoscopy allowed them to be self-made speakers and singers, the generational quality of much of the knowledge they would learn to possess also gave them access to less individualized notions of vocal culture and praxis.

This is not to say that (old) age was celebrated per se, but rather that men's claims to maturity intersected with their ability to interact with their physical environment in more controlled and fortuitous ways than women or children could. Throughout the nineteenth century, as notions of biological destiny and evolution would slowly come to replace ideas on environmental determinism, vocal experts found themselves in a quandary. As a physical phenomenon, wandering through space and traveling between bodies, the voice seemed particularly open to environmental influence and interference. After all, acoustics showed that the (spatial) environment coproduced sound, which implied that vocal sounds, too, were dependent on, for example, air quality and the built environment for their existence. On the other hand, being produced by the human body, the voice was now also to be included in new accounts of human biology, and those were increasingly written as stories of bodies that were not fluid, but self-contained entities, developed to perfection by evolution rather than constantly open to change under the more superficial influence of one's immediate environment. Stories of voices changing under the influence of the climate, like Giuditta Pasta's allegedly dropping into an alto range simply by moving to rainy England, persisted, but lost most of their plausibility throughout the century. The singer's and speaker's body could, in light of these ambiguities, only be imagined as partially permeable, and gender and maturity played an important role in modulating such permeability. Men, who were credited with more control over their bodies,[6] would be imagined as far less pliable or open to the environment (and as more evolved overall) and therefore in better control of their voices. The treacherous sounds of the pubescent male voice, however, made it all too clear that such stability was reached only with maturity.

In the second half of the century, as evolutionary interpretations of human biology gained ground, the traditional, almost dynastic approach to knowledge and authority no longer sufficed. Authors of singing manuals such as W. Hayle Walshe, William H. Lawton, Wesley Mills, and many others felt obliged to cite medical and scientific work as well.[7] Some, like Madame Seiler, even conducted their own research, publishing mixtures of protocols for original experiments, popularized accounts of medical and anatomical knowledge, and musical or artistic advice on vocal technique and style. As Seiler, who based much of her book on the "latest investigations and discoveries of Prof. Helmholtz in Heidelberg", noted, "for the artistic culture of the singing voice the knowledge of the physiological processes during the formation of tones does not suffice." The physical laws "relating to the same" were to be studied as well.[8] Like Behnke and Browne, many of these authors also attempted to trace their interest in vocal physiology back to the great Italian masters, thus modernizing their own discourse while strengthening their connection to the great (imagined) artistic past. Canto teacher William Shakespeare, for example, recognized that Italian masters such as Lamperti "had no acquaintance with anatomy," but nevertheless claimed that his singing method was based on both these old techniques and the modern, scientifically studied, principles of breathing and vocal registers.[9] At the same time, voice specialists such as Leo Kofler also drew attention to their own experiences as speakers and (semiprofessional) singers, thereby underscoring the growing level of cooperation and exchange in a stabilizing field. According to Kofler, "the study of physiology without the sound principles of the Old Italian School will enable a man to become a successful butcher, but not a trainer of voices."[10] Johannes Müller—famous for his experiments on dead larynxes—had claimed in 1839 that the singer's own somatic knowledge was useless for scientific research, because much like birds they sang without consciously thinking during the process: "they do it, without knowing much about how they do it." The physiologist was thus left to make his own, less corporeal and therefore more valuable observations.[11] Authors like Behnke and Browne, however, drew explicitly on the somatic experience of the vocal professionals they increasingly saw as colleagues: their publication on children's vocal education, for example, was entirely made up of questionnaires sent out to voice teachers and singers, and the (anecdotal) evidence gleaned from their answers.[12] Some authors, much like Garcia, staged themselves as the objects of their own experiments.

In scientific treatises and vocal manuals, a dizzying array of voices was brought to the reader's attention: not only was the pathological or the artistic voice opened up to scrutiny, but the author's voice was reflected upon as well, often as a yardstick for what was normal, intuitive, or recognizable to the reader. Moreover, in their professional lives, vocal specialists became vocal performers as well when lecturing upon their subject. Drawing attention to the corporeal character of voice production and vocal sounds therefore necessarily implicated the speaker's body as well: public lectures were opportunities to stage interesting, pathological or cured voices in an often theatrical way, in which the speaker participated not only as a bearer of knowledge, but also as an experienced vocalist. Well-known authors on sound and the voice such as Alexander Bell, Morrell Mackenzie, and Leo Kofler regularly lectured on their subjects (and attended the talks of their colleagues), and would therefore be able to claim firsthand knowledge of the difficulties and challenges of public speaking.[13] For them, too, the "beautiful speaking voice" would have been "a valuable asset."[14] As choir director Cyril Rootham noted, "for all lecturers, preachers, teachers and public speakers," a pleasant voice and good elocution were of equal importance.[15] Moreover, the speaker—who interpreted the experiments staged—also acted as a conduit between the audience and the voices on display. He represented the normal sound with which the scientifically informed audience could identify, or to which an audience struggling with speech impediments could aspire.

The assumed similarity between the expert's voice and that of his audience was replicated in writing as well, as authors appealed to their readers to empathically imagine their own somatic experiences. Stammering experts included emotionally charged descriptions of their own time as a "despondent and sorrowful being crushed by the burden of [their] infirmity" in manuals and self-help guides, forging an empathic relation between themselves and their (presumably stammering) audience.[16] The reader was thereby included in the vocal practices under scrutiny, sometimes even being spurred on to vocalize while reading. Educator Alfred Hoffmann did so, for example, when he asked the reader to "immediately" experiment with consonants. In his manual on vocal pathologies, aimed at teachers, he asked the reader to immediately attempt to distinguish between voiced and unvoiced phonemes first by aural trial and then by a test on the voice box. The trial has to extend to those consonants that are represented by the letters f, h, l, m, n, r, s, ch, sch, w, b, p, g, k, d, t.[17]

Like the experiment proposed by Behnke at the beginning of this chapter, Hoffmann's string of consonants is a call to immediate action: the letters are to be vocalized immediately to accompany and aid the act of reading.

In scientific manuals, some authors encouraged their readership of (potential) physicians to use their own vocal experiences—presumed to be healthy and normal—as a yardstick against which they could hold the pathological voices described in the manual and observed in their patients. In self-help manuals, in which the readers were presumed to be patients and healers at the same time, a similar technique was used. Readers were encouraged to vocalize while reading and thus learn to identify their own voices' pathologies. The ideal reader of these manuals would transform himself into a normal speaker by continued vocal practice. He would be guided by a book, and use his other somatic experiences (mainly hearing and touch) as a controlling device. By the time the patient reached the end of a manual, he would not only be able to assume the normal voice that was described in more formally scientific manuals, but would also be able to claim the vocal expertise that underlay their discourse. Or that was, at least, what more and more voice therapists started to claim. By the end of the century, being an expert on the voice coincided with vocal experience, either as a representative of vocal ideals or as a struggling vocalist.

This chapter traces the evolution of vocal expertise and experience from approximately 1860 to 1900, paying particular attention to the role of the physical and somatic practices of experts and their patients, readers, and pupils. The connection between consciously experienced and practiced vocal sounds and the personal experience of a body producing them led to a discourse in which both the voice and speech were seen as corporeal actions that could be controlled by the will. Thus produced, the voice was increasingly seen as a highly personal attribute. Although its production was understood as subject to universal natural laws, vocal practices were part of the performance of the modern self as they built a speaker's or singer's specific vocal identity.

The Expert's Body

The role of the scientist's body in empirical research has been an ambiguous one from approximately the seventeenth century.[18] Our image of the developing natural sciences is one of increasing advances in technology and methodology to ensure more objective and sophisticated measurements and observations, while the scientist's physical traits and experiences slowly disappear from view: the ideal observer, it seems, is only a mind—he is intangible, invisible, and does not meddle with the evidence through his own physicality.[19] However, while this general process of distancing the body from scientific observation was taking place, many scientists continued to use their own bodies as objects of observation and experimentation. Famous examples like neurologist Henry Head, who had his radial nerve severed, or the notorious father of LSD Albert Hoffmann show that the practice of using one's own somatic experiences never disappeared. However, this way of employing the scientist's body differed from one in which somatic experiences were used as means of observation. Rather, the somatically experiencing body became an object of observation itself: the scientists' somatic experiences were under scrutiny, and observed carefully (and often using machinery).

The production and potential pathologies of the voice were an obvious site for this type of corporeal involvement: unlike a lot of other physical ailments and deformities, the voice could not be seen, and therefore seemed to escape objective scientific observation. "For a long time, the ear was the only tool for analysis," physiologist Marage taught his students, "we were limited to a study of vibration by ear." This was highly problematic, because "not everyone has a good ear" and different observers could hear different things. "It is often said that of tastes and colours there is nothing to be disputed; but to discuss sounds is even more difficult."[20] Observers were reluctant to trust their hearing as the only source of information for diagnosis. Relying on one's own corporeal experiences as a singer and speaker allowed the inclusion of the other senses as well: touch, proprioception, and sight (as Garcia so exemplarily showed). Meanwhile, auscultation as a method of observation was developed, and made increasingly more objective hearing possible, as sounds could now be turned into visible symbols by disinterested machines.[21] Although somatic experiences gradually lost their authority as sources of knowledge, they seemed to gain importance as objects of observation. Moreover, in research on the voice in particular, the vocalizing, sensuous body of the scientist gained importance as a vessel for information and a means of sharing knowledge. Not only was his voice a carrier of discourse as he spread the word about his discoveries in classrooms and public lectures, his personal experiences as a speaker or singer could be related to the public as well. Presenting themselves as cases in their own research, vocal scientists managed to claim a role as modern, scientific observers while also maintaining a vested interest in the subject that would suggest a similarity between themselves and their pupils or patients. The somatically experiencing body of the scientist was not only an object of research, then, but also a site for sharing experiences with other speakers, singers, and stutterers, and therefore—through its intuitive recognizability—a sign of trustworthiness.

The development of a rapport between the scientist and his audience by the recognition of (vocal) similarities was perhaps most important in the staging of public lectures, which aimed to spread known knowledge rather than generate new research. As Paul Henry Gerber stated in the introduction to the publication of his *Seven Popular Lectures*, "it was not their goal to introduce new things, but rather to make old and known things more widely familiar."[22] University courses had a similar goal: Dr. Marage, who taught vocal physiology for singers at the Sorbonne, had his lectures recorded by his wife and then published them accompanied by a host of images detailing the experiments and machinery used to obtain the scientific results described, thereby possibly replicating the kind of theatrical performance audiences would expect from a public lecture on medical discoveries.[23] The manual contained, for example, a careful description, schematic rendition, and image of an "apparatus to photograph the voice," along with some resulting photographs and a rendition of a rather odd-looking "siren for vowels and oral resonators" to make the described scientific results transparent and concrete.[24] Public lectures were regularly enlivened by exhibiting live experiments showing off new machinery, including members of the audience (who were occasionally blindfolded or otherwise impaired[25]) in the performance, or inventively representing the inner workings of the body—for example by using bottles filled with fluid as stand-ins for the lungs. They also provided a theatrical way to showcase the human voice itself. The already-mentioned Lennox Browne used a group of singers, positioned behind a screen, to show the compass of the human voice and explain the concept of vocal registers:

> My friends who are here behind the screen, for a purpose that I will acquaint you with presently, will now sing from bass up to soprano, and you will hear at once what an extended compass the human voice has. . . . The next illustration I desire to give you is in relation to the difference between the male and the female voices . . . The tenor has a few notes lower than the contralto, who, on her part, possesses a few notes

> higher than the tenor; but within certain limits the voices are produced by the same mechanism, and, provided a contralto and a tenor are chosen of similar timbre of voice, it is often difficult to distinguish (their figures being concealed) whether a produced tone be that of the male or the female. (This experiment was made on the occasion, and until the audience was accustomed to the tones, there was considerable doubt in the direction indicated.)[26]

The experiment established his authority as an expert who not only was capable of fooling the audience—thus debunking popular ideas about vocal knowledge and replacing them with more legitimate scientific knowledge—but also showed how easily the sense of hearing could be manipulated, thereby stressing the need for other methods of observation. The experiment was, as Browne noted, explicitly designed to combat "those unscientific statements made by persons of educated or scientific pretensions which can be shown to be erroneous with hardly more than a moment's serious consideration."

Although Browne's staged experiment underscored the superiority of the expert's knowledge and distanced him from the crowd, the staged experiment worked because of an assumed empathy between expert and audience. Browne could easily predict his audience's faulty observations because he expected their hearing to be exactly like his own. Likewise, other vocal experts drew on the assumption that their audience's voice would resemble theirs. Nevertheless, their role in the performance of a public lecture opened them to specific scrutiny. Rather than self-presenting as the disinterested, incorporeal, almost invisible observer, the scientist giving a public lecture was making himself visible and audible to all.[27] When they presented themselves as experts on the voice, it is likely that their audiences would expect them to be expert vocalists as well, showcasing exemplary breathing technique, faultless articulation, and perhaps even a good singing voice. Much like other scientists did in laboratories, vocal experts depicted themselves as cases of vocal practice onstage—mostly as typical cases, representing the assumed normal human voice shared by the (majority of the) audience, but occasionally also as an example of a pathological voice performing its own cure. In an 1861 lecture on phonation, for example, Charles Bataille noted that he was presenting with the authority of someone who had "experimented much on myself."[28]

Around the turn of the century, Benjamin Beasley embarked on a series of lectures.[29] Because he had been a stammerer himself, his method for curing a stammer largely drew its authority from Beasley's status as a "fellow sufferer" and a man of great personal experience. Somatic practices and empathy were at the heart of his public image, rather than systematic and disinterested observation. Reports of his lectures show that he encouraged his audience to assume a similarly empathic position: rather than absorbing the expert's discourse, audiences were to trust their senses (mainly that of hearing), and to draw their conclusions about his method by observing the performed proof of its effectiveness. One audience member reported, "It was almost marvellous that Mr. Beasley, after having been himself a terrible sufferer for many years, should now be able to speak with so much freedom and oratorical power. He was a living instance of the success of his own system."[30]

In observing Beasley's personal vocal performances, these audiences replicated the actions of visitors to the therapist's institution, where regular "speech days" were held to convince interested (potential) patrons of the effectiveness of his method. The *Birmingham Daily Gazette* reported rather cheerfully about the speech days held in the summer of 1883: "The success of Mr. Beasley's system was thus beyond a doubt practically demonstrated, and as the relief afforded is not temporary, but permanent."[31] Beasley's voice, as it was staged, was just one of many successful examples of a cured stammerer, a case study exemplifying the reliability of his own methods.

The practice of staging one's own voice as a kind of case study for the audience was part of an educational practice as well. Canto teachers had already been using their own voices as examples for their pupils, but in the nineteenth century the practice grew in scale as it was replicated on paper. The singing methods of grand opera stars like Enrico Caruso, Sims Reeves, or Mathilde Marchesi banked on the celebrity status of their authors (because who would not be curious about the lives and routines of these heroes of the stage?), but less illustrious authors were equally generous with descriptions of their own practices of voice cultivation.[32] Teachers of singing and elocution committed to paper their corporeal approaches to teaching, and so did empirical scientists. Leo Kofler, for example, included long autobiographical notes in his *Art of Breathing for Tone Production*. Billed as a "A book indispensable to singers, elocutionists, educators, barristers, clergymen, and to all others desirous of having a pleasant voice and good health," it focused on its author's musical training and health for several chapters. "I shall have to apologize to you, dear reader, for telling you more of my early musical life than I promised," he noted after a couple of pages, but went on to describe his laryngeal problems and the development of his breathing for several more.[33] Stars of the operatic stage, who used their fame as a platform to sell singing manuals, could present their voices as exemplary instances of vocalization that should be studied in order to be imitated or merely admired from afar. Caruso, for example, claimed that he had "been blessed with what is called a naturally placed voice, and never had trouble with the mezza voce. The majority of Italian singers come to it easily."[34] Scientists, therapists, and canto teachers who did not have an illustrious stage career, on the other hand, presented their own voices as case studies with which the reader could identify. The individual voice of the author appeared as a salient example among other cases, and was similarly opened up for observation. This seems to have been common practice in training as well, as a booklet by Emil Behnke and Lennox Browne on the training of children's voices shows: the publication contains the opinions of numerous educators on young voices, but it is interlaced with the teachers' reflections on their own health, their history as singers, and the state of their own voices.[35]

The advantage of including one's own voice as a case study was that it could be observed more carefully and completely. It was always available for any experiment, and all the observer's senses could be engaged. In 1839, Johannes Müller had stated that the singer's own experience of his voice was essentially useless for scientific research, but by the 1860s, educational, elocutionary, and physiological specialists were drawing on one another's work in different ways. The method of self-reflection and somatic observation that had been used by elocutionists also started to gain a (small) place in empirical work—where vision seemed to be reaching its limits as a means of objective observation, and the mechanization and systematization of all somatically observable phenomena started to replace the eye. Machines and devices like tonographs or eidophones went beyond seeing the voice as it was produced by human and animal anatomy, but allowed scientists to study sound per se.[36]

Moving away from the purely visual and anatomical did not dislodge the voice from the body, however. If anything, the stress on the somatic engagement with the voice anchored its production, control, and perception firmly in a highly individual body. "Voices differ," as Henry Wood noted, "as individuals differ."[37] Others compared the voice to the face and its expressions. According to Theodore Schmauk, "timbre is the distinctive blossoming of the individual personality. There are no two faces alike. Still less are there two voices alike."[38] Like the face, the voice was individual, and capable of expressing deeply felt emotions: "Gladness, sadness, confidence, shrinking, boisterousness, harshness, coarseness, peevishness, ostentation, and higher virtues and lower vices leave a more and more permanent impression not only upon the face, but upon the voice."[39] It also expressed temperament and personality. Rather than a given mechanism, governed solely by natural laws and produced by predictable movements, the voice was increasingly understood as the result of a number of corporeal, emotional, and above all personal characteristics. The cooperation of physiologists, elocutionists, and educators thus went hand in hand with a broad, almost holistic view of the body producing the voice. In their cross-disciplinary work, individual cases and patients were described with such accuracy that the reader could almost feel he knew them.

It is perhaps no surprise that authors who valued their own somatic experiences also engaged with the personal stories and sensations of others. If expertise and experience were indeed related, they would have shared their authority at least partly with others. Even though physiological and elocution manuals are highly normative texts, imparting a specialists' knowledge to their readership, many manuals seemed to assume the tone of the increasingly popular method of object education. The reader was to experience (see, hear, feel) the phenomena described, and would learn by practice rather than by example. Various exercises for breathing, reading aloud, and vocalizing accompanied the more abstract explanations of vocal practice and health in many manuals.[40]

Instead of a return to an older paradigm in which patients derived agency from their exclusive knowledge of their own invisible symptoms, this approach to the individual voice (including one's own) as a case pointed to an increasingly sophisticated image of universality among the human species.[41] The most eloquent description of the way in which a case could be approached within this universal view of man was one by Swiss educator Heinrich Pestalozzi. Pestalozzi was an active and well-known propagator of the method of intuitive education, and, as a small 1910 publication attests, a keen amateur vocalist as well. In the introduction to *Avoiding Mistakes in the Formation of the Human Voice Through Individual Vocal Education*, he reported on experiences in "his own throat": "I often sang for a long time in this or another way, until bad experience drove me to try something else. This way, the natural laws gradually revealed themselves to me, and I learned to single out the right method among many different ones. I am therefore confident, that every person who strives and researches honestly will come to the same conclusions in a naturally and very individual way."[42]

The work was conceived almost as an object lesson: Pestalozzi assured the reader not only that his individual experiences were relevant, but also that the reader's experience would be equally valuable. He then proceeded to painstakingly describe his own evolution as a vocal student, as well as that of a handful of others. "Fräulein B.," "Fräulein C.," "Herr D.," and "Herr E.," each representing a vocal type, were presented to the reader. The reader, present in the book as the elusive "Herr A.," was encouraged to use his own body as a site of experience and experiment. According to Pestalozzi, the good teacher would show healthy vocalization "am eigenen Leibe" (with his own body), and the pupil "Herr A." "can observe in himself" what the effects of the technique would be.[43] The reader of the booklet was presumed to be "searching and struggling honestly," was free to compare himself to the cases, or between the cases, and extrapolate from these to learn about his own voice.

While this intuitive method seems to be a far cry from the statistics that were starting to be produced on the voice in the 1830s (and that continued to be compiled and published throughout the century), it seems to have aimed for similar results. Pestalozzi's style may have looked free and natural, but it remained educational as well. It was a surprise to no one that "Fräulein B.," the alto, was described as possessing "colossal vocal means. Her voice intones voluminously but dully."[44] The soprano's voice, on the other hand, was "pretty" but sounded a bit nasal. That the tenor had a small but brilliant voice, and the baritone a powerful voice—albeit an unfortunate tendency to shout—was no coincidence either. These rather prescriptive notions of different voice types are indeed very similar to the disciplinary vocal framework produced by schematic and statistical representations of the human voice by physiologists, such as J. Müller's table of the compass of the voice in "different individuals" (read: singers of different gender and voice type).[45] By the 1860s, this type of representation of vocal registers had become ubiquitous. It was laid out by scientists like Édouard Fournié, Morrell Mackenzie, Marage, and Frederick Mott, who were involved with careful measurements of larynxes, vocal cords, and pitch (Mott even expressed the compass of the voice in "vibrations per second," which heightened the image of mathematical precision).[46] Teachers of canto and elocution relied on them as well. In their work, the experimental origin of the graphs was usually only obliquely referred to. Henry Holbrook Curtis, for example, who included a diagram of the vocal compass in his book on *Voice Building and Tone Placing*, paid lip service to the work of Mackenzie, Behnke, Garcia, and Seiler—thereby possibly underscoring his own scientific credentials—but did not go into any detailed description of their work to legitimize his diagram.[47] In some cases, research or theories underlying the distinction between voice types became invisible altogether. James Farrar's *Practical Book for Orators, Clergymen, Vocalists and Others* showed the compass of the voice quite precisely, using musical notation to show the tessitura of each voice type, but provided no explanation or background for these distinctions.[48]

If, from roughly the 1830s up to the 1860s, the compilation of statistics and graphs was mainly a way to accumulate information and share it with one's fellow scientists, the later graphic representations of vocal difference seemed to serve a different purpose. They allowed scientists to spread knowledge gained by empirical research to a large audience, and to introduce them to a discourse in which vocal performances could be approached systematically. Especially representations that used musical notation (rather than actual statistics) allowed readers from other disciplines to absorb the physiologist's conclusions at a glance: rather than providing means and percentages (as was the case, e.g., in the statistics that were compiled on the geographical and social distribution of stammering), these graphs claimed to show "normality" and thus produced a framework in which each reader was invited to position himself. A soprano reading Pestalozzi's booklet on the avoidance of vocal mistakes might well have wondered if she, perhaps, was producing tones that were rather too nasal as well. Referring to schematics like those of Farrar's would allow her to establish whether she actually was a soprano in the first place. For those who failed to fall neatly within the described or prescribed registers, manuals to train their voices into normality were manifold. Or, conversely, readers could diagnose themselves with an unconventional or even pathological voice.

The Reader's Body

In their attempts to reflect on different bodies and somatic experiences, authors also activated their audience's bodies and senses. Especially in the more popularized work—which would presumably have been read by an audience who identified with either vocal skill or vocal pathology—an active, corporeal engagement with the text was encouraged. The already-mentioned use of simplified schematic renditions of statistical material was the most popular example—and possibly the most authoritative one as well, as it suggested mathematical precision and scientific proof. The difference between male and female voices, for example, was presented as an unchangeable and undoubtedly natural distinction. In the first half of the century, too, authors noted a pythagorically pleasing distance of an octave between male and female equivalents of voice types. According to J. Müller, for example,

> The compass of the voice in different individuals comprehends one, two, or three octaves; in singers,—that is, in persons capable of singing—it extends to two or three octaves. But the male and female voices commence and end at different points of the scale. If we commence the scale at the C of the eight-feet organ-pipe, the male bass voice will take its

> beginning from the first E, the barytone at A, the tenor at C1, and these male voices will reach the bass to A1, or higher, the barytone to F2, and the tenor to C3. The female voice is as deep as the male only in masculine women. The voices of women, boys, and eunuchs begin between F1 for the alto, and C2 for the soprano and reach as high as F3 for the alto voice, or A3 for the mezzo-soprano, or C4 for the soprano at the furthest as high as F4. The lowest note of the female voice is therefore about an octave higher than the lowest of the male voice; the highest note of the female voice about an octave higher than the highest of the male.[49]

In the second half of the century, this categorization was confirmed in repeated experiments and statistics, and emphatically demonstrated by neatly drawn diagrams.[50] Some of them even detailed what operatic roles would be suitable for which compass of the voice. Illustrations could be similarly used by readers as a point of reference for their own throats, bodies, and posture. Giving readers the possibility of imagining vocal normality thus also subtly suggested the necessity to embody it—the disciplinary tone of good breathing, natural sounds, and pleasant voices was by no means limited to educational or lay manuals. In medical handbooks and scientific essays, a similar terminology of vocal propriety and esthetics was used.

Especially in these texts, the author and his body provided an important bridge between subject and reader: as the author positioned his own voice within (or outside) the parameters of normality, the reader was expected to empathize and even identify with the author. In medical manuals, this empathic recognition was based on shared expertise; in work by former stammerers, on shared suffering; but on a broad scale it could also be based simply on shared humanity, as many of the vocal experiences and practices described were thought of as universal—as part of what was understood as the normal human body. This universalist streak led to multiple ways in which the reader's body in manuals and scientific treatises could be activated. In the first place, they could be called upon as an available test site for the author's pronouncements. Readers were encouraged to try to feel or see the anatomical and physiological phenomena they read about by observing their own bodies. This allowed them, to a certain extent, to share authority with the author and exert control over their own knowledge. Second, the discursive activation of the reader's body was aimed to afford them control over their own voices as well. As the understanding of voice and vocalization shifted from a purely anatomical, mechanical process to an individualized, somatic one, the speaker's personal will became an object of concern. Readers were encouraged to willfully control their voices—especially in educational books and self-help manuals—and control over the body, its organs and its muscles, was presented as a way to establish that control.

The issues of control all seemed to hinge upon detailed knowledge of the body—the human body in general, and one's own in particular. Throughout the nineteenth century, authors continued to provide more or less detailed anatomical drawings of the larynx and the vocal cords. From the 1860s onward, these were often accompanied by an image of the throat as seen through the laryngoscope. These images did not really offer more insights into the makeup of the throat, but provided an image of the voice at work, thus stressing the importance of various muscles, of the will, and of performance in general for the production of the voice. Being able to see the voice during vocalization bore the promise of making its mechanisms and workings more transparent, and therefore more meaningful. According to George Duncan Gibb, who was an early adopter of the new instrument, "In the hands of those accustomed to its manipulation, its use effects what has been heretofore looked upon as marvelous, for it [the laryngoscope] renders 'the dumb to speak and the deaf to hear.'"[51]

Images were shown of generic normal and pathological throats, which could be compared to one another and to those of patients. Moreover, after the example of the pioneer of laryngoscopy, medical students could examine their own throats—and were admonished to do so as a kind of practice before subjecting their patients to their undeveloped skills. Inserting the laryngeal mirror in their own throats would familiarize them with its (often unpleasant) physical effects.[52] Manuals for a less specialized audience relied on less sophisticated methods to examine the throat. Alfred Hoffmann, who explicitly aimed his work at educators, suggested to his readers in *Pathologies of Speech and Voice* that they could lightly press three fingers on their larynx, or press their fingers into their ears, to experience the vibrations of their voice in their bodies.[53] The goal of this kind of exercise was twofold. On the one hand, the somatic experience of vocalization could help readers to picture the inner workings of the throat: it taught them that vibration was, in fact, connected to sound, and *where* this sound was produced.[54] On the other hand, it also familiarized them with the workings of their own, presumably normal but nevertheless particular, throat. Rather than merely imagining the feelings of others (such as the author's, or those of described cases and anonymous larynxes), these practices heightened the sense of empathy between author and reader as they presumably shared the exact same experience. The author would feel the vibrations of his own throat when devising the little experiment, and the audience would feel compelled to do the same while reading (it is indeed hard not to grab at one's throat while reading these manuals). Much like an audience at a public lecture may have synchronized their experiences with that of a speaker showing vocal practices on stage, readers would have co-experience with the author.

This type of DIY experiment also provided the reader with a mechanism of control: he could test the assertions of the author (as he felt his throat trembling under his fingers while vocalizing, the description of voiced consonants as vibrations appeared proven). Singing teacher Wesley Mills even explicitly presented some simple experiments to "convince those who may not have given much attention to the subject that sound is due to movements of some object, which we term the sounding body, strictly that which starts the vibrations by its own movements or vibrations."[55]

Moreover, discerning readers could use these and similar somatic techniques to discipline and control their own vocal practices. This was particularly salient for an audience hoping to learn something from the expert discourse they were buying and reading. After having diagnosed their pathological voice, or identified the character of their singing voice ("I am a tenor"), they could employ their hands, eyes, and ears to establish what they were doing right—or wrong. More than anything, this use of manuals would rely on the readers' will: not only would they have consciously bought a manual to improve and alter something as supposedly natural as their voices, most manuals actually promised the possibility of indeed willfully doing so. While introductory chapters on the working of the human voice suggested the universality of the mechanism, and more narrative prefaces stressed the utter individuality of each voice, the bulk of each manual was geared toward an understanding of perfectibility that was guided by universal principles, but ultimately rested on the individual will to engage the body in the search for the ideal voice.

Many authors played on the intuitive understanding that weak voices were the result of weak personalities, with overwrought emotions and frayed nerves, and therefore suggested that the mere ability to apply oneself to self-improvement would result in stronger vocal performances. George Gibb noted that adolescents, especially, were prone to the kind of vocal weakness that essentially expressed a more general nervous problem, referring to it as a "constitutional weakness of some portion of the spinal nervous system, especially implicating the throat and chest, for the intellect is perfect, although the listlessness occasionally present would leave the impression that it was not clear."[56]

Rather than setting the bar for a particular sound, then, the ideal voice was described in emotional instead of esthetic or acoustic terms: the healthy voice exuded a sound of control, because its owner was, physically, mentally, and emotionally, in control of its production. These levels of mental, emotional, and corporeal control almost seemed interchangeable, as physical exercise was thought to produce steadfastness, while mental stability enabled one to cultivate the discipline necessary for the control of organs and muscles.

The most obvious references to the importance of the individual will, and its corporeal manifestations, were directed at stammerers. Because their speech was characterized by such a noticeable lack of control, the cultivation of a willfully controlled body to reign in a stammering tongue or garbled speech was a powerful discourse. The origin and role of the individual will in exerting that control changed throughout the century—moving from an emotionally charged understanding of the stammerer's nervousness to an exceedingly technical and electrical understanding of his overwrought brain. However, techniques to develop the will and subject the stammering body to its control hardly changed at all. They relied on various senses (seeing, hearing, proprioception) that would allow the stammerer to monitor his own body and—eventually—his speech.

Regulating the stammering body and its syncopated movements was part of an earlier arsenal of stammering cures. Objects to aid stammerers in their cure, often likened to the pebbles Demosthenes was believed to have held in his mouth, all served as external means of control. The *bride langue* was inserted in the mouth to depress the tongue, the *muthonome* regulated the rhythm of speech, and so on. Although these mechanical crutches did not disappear outright, they were increasingly dismissed as quackery.[57] Rather than relying on external regulators, the late nineteenth-century stammerer was to gain complete control over his own body. His own ears alerted him to his own faulty speech, but also allowed him to listen to others for examples of good speaking. His eyes observed actors speaking onstage, after which he was to model his own speech by speaking in front of a mirror, "for the purpose of showing the reader the exact manipulation of the articulating organism. It is also to encourage him to note daily, the improvement he has acquired in lip and jaw flexibility; and to notice the increased ease of his articulation, by the practice of these varied and indispensable muscular movements."[58] And, finally, he alone was held responsible for the cultivation of a healthy, strong body capable of perfect muscular control. By the end of the century, stammerers were their own experts, and figures like Benjamin Beasley proudly claimed to have cured themselves through sheer perseverance. This shifted the narrative: rather than seeing stammering as the result of emotional or moral weakness, stammerers were framed as courageously battling their difficulties. Beasley put this reversal of the connection between

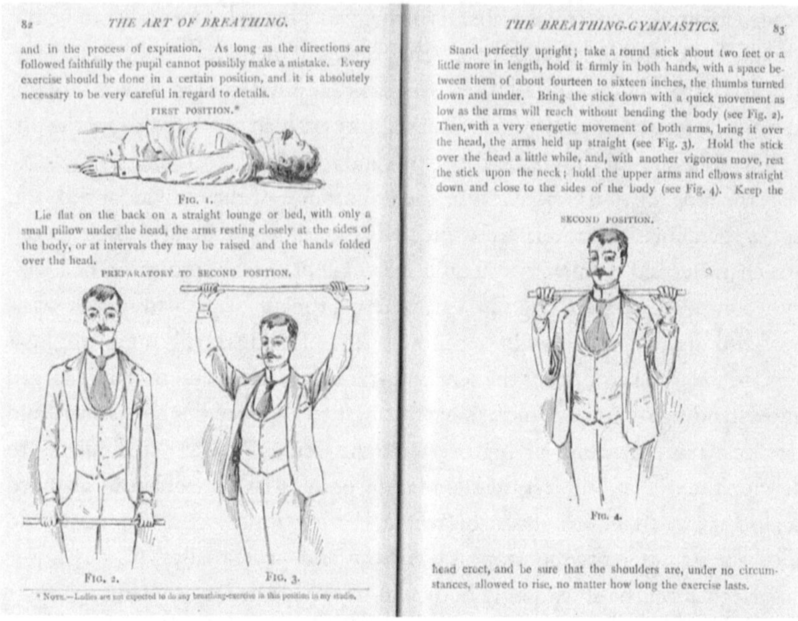

82 *THE ART OF BREATHING.*

and in the process of expiration. As long as the directions are followed faithfully the pupil cannot possibly make a mistake. Every exercise should be done in a certain position, and it is absolutely necessary to be very careful in regard to details.

FIRST POSITION.*

FIG. 1.

Lie flat on the back on a straight lounge or bed, with only a small pillow under the head, the arms resting closely at the sides of the body, or at intervals they may be raised and the hands folded over the head.

PREPARATORY TO SECOND POSITION.

FIG. 2. FIG. 3.

* NOTE.—Ladies are not expected to do any breathing-exercise in this position in my studio.

THE BREATHING-GYMNASTICS. 83

Stand perfectly upright; take a round stick about two feet or a little more in length, hold it firmly in both hands, with a space between them of about fourteen to sixteen inches, the thumbs turned down and under. Bring the stick down with a quick movement as low as the arms will reach without bending the body (see Fig. 2). Then, with a very energetic movement of both arms, bring it over the head, the arms held up straight (see Fig. 3). Hold the stick over the head a little while, and, with another vigorous move, rest the stick upon the neck; hold the upper arms and elbows straight down and close to the sides of the body (see Fig. 4). Keep the

SECOND POSITION.

FIG. 4.

head erect, and be sure that the shoulders are, under no circumstances, allowed to rise, no matter how long the exercise lasts.

Figure 6. Leo Kofler, *The Art of Breathing as the Bases of Tone-Production: A book indispensable to singers, elocutionists, educators, barristers, clergymen, and to all others desirous of having a pleasant voice and good health*, London: Curwen and Sons, 1901, pp. 82–83. Courtesy of the Wellcome Collection.

stammering and nervousness very clearly: "Nervousness exercises a very predominant influence over stammerers, but it is not, as many suppose, the cause of stammering. Stammering is the cause of nervousness. . . . My own personal experience unquestionably proves to me how fearfully one may be afflicted with the impediment without having reason to attribute it to nervousness. As a child I was very healthy and robust."[59] The imperative for willful control of the voice through the body was not limited to pathological voices, and could take much more specific forms as well. The lungs, especially, were targeted as an area of vocal cultivation. Although, as many authors pointed out, the lungs were not muscles and could therefore not be directly steered, most manuals suggested that it was possible to train the lungs—and thus subject them to the will—on a long-term basis through breathing exercises. Leo Kofler devoted a whole book to the *Art of Breathing as the Basis of Tone Production* and was convinced that his breathing exercises, because of their clear benefit to health, were a better life insurance than "the small policy in the New York Life Insurance Co."[60] Because the voice was produced by pushing air through the larynx, healthy lungs were thought of as the basis for healthy speech and, as Kofler's volume demonstrates, faulty vocalization was linked to faulty respiration. For women, the constraining effects of the corset were often mentioned as an important cause of imperfect breathing. In 1884, Lennox Brown expressed the hope that "some success is at last resulting from all that has been written, and that those at least who value their voices and their health more than what they are pleased to call their figures, are awakening to the fact that a beautiful woman need not necessarily be a deformed one."[61] Rather than clavicular breathing, voice trainers and physiologists alike suggested abdominal breathing and the removal of the corset.

Men, who were not confined by the corset, were even more often admonished to train abdominal breathing. This technique would not only allow them to develop a powerful voice, it also marked them out as men per se. Emil Behnke was convinced that one could tell the difference between male and female singers merely by watching their abdomens:

> By way of illustrating this curious difference of breathing in men and women, the following anecdote, which has the recommendation of being strictly true, may perhaps amuse the reader. Some time ago a troupe of "Female Minstrels," calling themselves, I believe, "the American Amazons," made a tour through this country. Their faces were blackened in the orthodox fashion, and they were in male attire, wearing tight-fitting garments of a peculiar kind. Two friends, both medical men, went to hear them (or perhaps to see them, I am not sure which), when Mr. A remarked that two of the performers were men. Mr. B did not see it, even when the individuals were pointed out to him, and asked his friend for the reasons for his opinion. "Why" said Mr. A, "I see it by their abdominal breathing!" And sure enough Mr. B now saw it too, and there was no mistake about it; for in the two suspected individuals the abdomen was evidently moving in respiration, while in all the others no movement was perceptible excepting that of their chests.[62]

Moreover, for men, regular breathing exercises had the added benefit of helping them "to acquire the necessary chest development."[63] The cultivation of the chest—enlarging its circumference as well as lung capacity—seemed to be an almost obsessive preoccupation of therapists and pedagogues of the period, and reactions to male corset-wearers were even more visceral than to their female counterparts.[64] It is clear, then, that not all clearly defined chests were valued equally: the exaggerated proportions of the corseted dandy were brushed off as effeminate, while the overly muscular chest of the body builder would only gain popularity much later, around the turn of the century.[65] Nevertheless, despite the repeated stress on the natural development of the male chest, esthetic criteria for a visually pleasing body were impossible to separate from arguments linking the well-developed lungs inside the virile chest to powerful, authoritative voices.

Likewise, a strong and well-defined musculature was easily associated with a supremely controlled voice, as both called to mind the physical strength and self-discipline necessary for their production. Like breathing exercises, gymnastic training was suggested by authors from all disciplines, and differentiated according to the type of body and voice considered proper for different people. Emil and Kate Behnke detailed a number of exercises in their book on the preservation and development of the speaking voice, supplying photographic images of a man and a woman performing them. Although the descriptions of the exercises were not explicitly gendered, the choice of photographic subjects shows a prevalence of suppleness and flexibility for women, and the development of upper-arm strength and posture for men. This differentiation dovetailed neatly with the expectations for male and female voices: valuing strength in the former, and elegance and flexibility in the latter.

Similarly, some authors differentiated between vocalists of different ages: children's constitutions were not to be overexerted, young men's bodies needed to be cultivated, and older men would have to rely on diligence in their youth to retain their physical and vocal strength. For girls and women, less differentiation is evident: as female voices were generally thought to go through less pronounced changes throughout their lives (their voices did not "break"), and were generally likened to those of children, the principles of prudence and limiting exertion were perhaps extended to them as well. After all, they were considered to be the weaker sex, with sweet and flexible, rather than strong or powerful, voices. Moreover, the (semi)professional path of moving from schoolboy to political (or otherwise professional) orator, or from choirboy to soloist, was not open to them, and long vocal careers did not need to be set out for girls. Finally, women were considered to be more natural vocalists. Therefore, they did not need to build up the same degree of physical resilience to buttress their voices. Richard Faulkner, referring to the research of the French Dr. Marage, offered a curiously precise narrative about the relative ease with which women spoke, arguing that they, like children, exerted less energy with their smaller vocal organs:

> During ordinary conversation, there is developed, in one hour, an energy of about 48 kilogrammeters, that is to say, that to speak during one hour is not more fatiguing than to lift at each second a weight of 13 grammes, one meter high: a lady in trifling with her fan, or a professor gesticulating with a piece of chalk, exerts an energy much greater. To pronounce a discourse in a large room, the energy is more considerable, but it is not enormous; it is, on an average of 200 kilogrammeters per hour; an employee of a railroad performs a greater work in lifting from the ground on to his shoulder, four packages of 50 kilogrammes. We have compared subsequently the energy developed in conversing by the voice of a man and that of a woman, and have found that women are fatigued, in talking, four times less than a man. We can then comprehend how children who have a larynx narrower than their mothers, can talk for several hours without taking a rest.[66]

The use of breathing exercises and gymnastics for the cure and strengthening of the voice was not strictly new in the 1860s. Neither was the self-help method.[67] Nevertheless, the second half of the nineteenth century saw a rise in this type of publication, which seemed to fit so nicely with the bourgeois ideal of the self-sufficient, self-made man, or the young lady determined to improve herself.[68] Moreover, the DIY therapeutics described in these vocal self-help publications gained legitimacy in wide circles of experts. Whereas, on many medical terrains, physicians were establishing their authority and claiming exclusive expertise over their patients' bodies, on the issue of the voice they seemed increasingly willing to share their status as experts with elocutionists, singers, and others, and with the patients themselves.[69] Key to this particular approach to vocal problems was, perhaps, that vocal pathology and vocal refinement were so deeply entrenched in the upper and middle classes. Rather than curing these diseases through a disciplining and civilizing framework (as was the case for afflictions like tuberculosis and rickets), physicians and elocutionists alike sought to strengthen the vocalists' control over their own bodies, thus allowing them to individually cultivate their voices.

Although this process rested on the practices of changing the body—and thus assumed its malleability—it mostly depended on increasingly fluid definitions of the voice itself. As the connection of the voice to individuality and will was strengthened, the stress on the performative character of the voice rose as well. Rather than being produced by a mechanism that could be altered only through surgical intervention, the voice was increasingly seen as a product of a particular person's body, personality, emotions, and (controlled) vocal practices. Expertise on the voice could therefore no longer rely solely on disinterested observation of the tangible parts of the organs producing it, but also had to take the somatic experiences of vocalists into account, turning every speaker and singer into a potential expert on their own voice—provided they had the ability to educate themselves and systematize their experiences. As modern, self-controlling individuals were gathering knowledge on their voices, and willfully cultivating their sound to adhere to social and esthetic expectations, voices became more and more consciously used as symbols of their owners' identity. Not only were there "as many voices as there were faces," but each voice also communicated as much about a speaker's gender, age, class, or profession as his clothes or gait could.

Modern Vocal Identities

The connection between the (acoustic) qualities of the voice and the performance of a modern identity was strongest for those who were professional voice users (singers, barristers, teachers, etc.) or for those who suffered from vocal impediments. As stated previously, the categories frequently overlapped: the conscious engagement with vocal practices, their social implications, and their professional effects were mainly the terrain of the professional classes. These were not only the people who would, practically, benefit most from a conventional and proper vocal sound, but also those whose social identity was most obviously intertwined with notions of control and self-determination. Those whose metaphorical voices were most implicated in the process of democratic politics—those whose rational speech was most valued—were also the ones whose voices were subject to scrutiny, and vulnerable to failure.

The whole concept of the individual voice as it was defined and described in physiological and elocutionist manuals was therefore already implicated with notions of gender, age, and class. Not only was the mature, male, and middle-class voice usually taken as the norm (or indeed as the only one worth researching), the way in which autonomy, individuality, and willful control were connected to the cultivation of the proper or natural modern voice suggested a connection with contemporary notions of middle-class masculinity. Here, too, an assumed solidarity or empathy between authors and audience was at play: the overwhelming majority of work on the voice was written by men, for men. Singing manuals were occasionally written by female soloists and/or directed at young ladies—and of course, women were objects of attention and study—but generally authors could quite safely imagine their readers to be a fairly homogenous group. Knowing winks toward an audience imagined as heterosexual and male, like this one in a stammering manual, bore witness to the fact that while the adult man was seldom clearly identified, he was implied to be central:

> One remark ought neither to be omitted nor disregarded, as it respects that interesting and important section of the community on whose character and manner the general well-being and happiness materially depend. I shall, of course, be immediately understood to allude to that fairest of Nature's works, "lovely woman." For can a bosom exist that would not throb with increased pleasure to promote the chaste and gentle influence of female charms? That would not gladly smooth away any rugged or unseemly blemishes which would weaken their bland and benign sway over our affections?[70]

However, despite the focus on mature male voices, and despite the rising importance given to somatic experience to claim expertise on the voice, other voices and the identities associated with them were included in manuals and treatises as well. From the beginning of the century, differently sized larynxes and their different sounds had been connected to the age and gender of the bodies in which they were lodged. In the latter half of the century, the differentiation of voices along lines of age, class, and gender were further developed. These differentiations went far beyond measurable units such as pitch and the size of larynxes or vocal cords, to also include vocal timbre, personality traits, and social accomplishment. The body worked as a focal point on which different sociocultural and vocal expectations could be pinned, and which could work associatively as a sign of the invisible qualities of voice and temperament. As Theodore Schmauk postulated,

> the physical and mental divergency between man and woman impresses itself upon the voice, entirely apart from the differences in pitch or power. Whether tenor or bass, and strong or feeble, whether alto or soprano, we recognize in them the difference in sex. There is an occasional anomaly. We sometimes say of a man "he has a woman's voice." And frequently the prevailing manliness or womanliness in the spiritual nature of one who is of the opposite sex, greatly influences the voice. But this is the exception. In listening to the Wagnerian singers, the manly depth, volume and

> nobility of Herr Scaria's bass was impressive. But the tenor or rather baritone of Herr Winkelmann was not any the less manly.[71]

Reframing popular clichés about various types of vocalists and couching them in "scientific" terms, the process of systematizing vocal differences not only shows how vocal science changed and grappled with the changing methodologies and notions of expertise and experience, it also gives us an insight into conventional social associations connected to different vocal sounds in the second half of the nineteenth century.

One of these conventional associations was the connection between (young) children and a lack of identity: as children grew up they were thought to not only acquire speech, but also develop a voice as they became a full-fledged person. There was some speculation on the humanity of so-called savages and deaf persons, and children often served as a means of comparison. The assumed infantile characteristics of these groups seemingly showed their lack of development. Although some vocal pedagogues (notably those involved with boys' choirs) were interested in children's voices in their own right, the education of the child's voice was generally seen as a preparation for the eventual cultivation of their adult voices, and extensive attention went into how treble voices matured and mutated into tenors, baritones, and basses particularly. Developing the voice at a young age was thought to be beneficial precisely because children were "unfinished" personally and vocally, which meant that their voices and personalities could still be molded. Moreover, children were thought to be particularly good vocalists, as they still spoke and sang naturally, without too much interference from civilization. As choirmaster and educator Kurt Schilling noted, for example, the notion that children were gifted with a naturally excellent voice was widespread, but experts like himself knew that early training, and not mere nature, was at the root of children's sounds. Letting children sing naturally ("wie ihnen der Schnabel gewachsen ist") would not suffice. Nevertheless, he advised teachers to select some exemplary pupils to provide their peers with examples to imitate, rather than to provide such examples in the teacher's own bass voice, which would encourage children to acquire bad habits: "Spare your voice, you keep complaining, and let the children set the example! This tone is much better suited for them than your deep bass tones, which only tempts them to push their voices, or your laborious attempts to sing high notes, which the children copy so wonderfully."[72]

What the educator needed to do, then, was above all to steer young singers toward the more genteel forms of nature, occasionally shaving off the rough edges of the most savage child singers (i.e., those of the lower classes). As British choir master Cyril Rootham pointed out, in agreement with his German colleague,

> The voice of the newspaper-boy in our streets is "natural" enough; for surely no classes exist to train boys' voices for calling the titles and contents of newspapers. Similarly, children at their games use their voices "naturally"; but all voice-trainers know the damage that frequently results to both the speaking voices and the singing voices of children who are allowed to shout constantly and without restraint. The truth is that singing is an art, even for children, who may, and should, be carefully taught to speak and sing properly from quite an early age. If the teachers be competent, there need be no fear of any taint of artificiality.[73]

As the voice could be developed and built, it could also go into decline and become lost. Much like childhood, old age was treated in most manuals as an exceptional period of transition in which special care had to be taken to position oneself opposite the normal state of vocal and physical maturity. The old voice was depicted as a crumbling building, and as a sign of the loss of gendered identity. As men and women aged, their voices were said to become increasingly indistinguishable, with women's pitches dropping and men's rising. The loss of a voice evidently also inaugurated a loss of individuality and identity. To counteract this, singers particularly were admonished to attempt to hold on to a youthful body and a great measure of control—as Schmauk noted, "age may conquer this tendency to deterioration. Regular practice and careful living will preserve the voice, as well as any other organ."[74] The issue of physical endurance and strength in particular was stressed in the context of upholding a convincingly masculine voice.

In the context of this linear image of the voice that came into being through cultivation, then reached its apex in maturity before going into decline, Shakespeare's eloquent description of the male voice descending back into childish helplessness in old age in *As You Like It* was deemed a relevant metaphor even in the incongruous context of laryngoscopy.[75] The poetic reference to the big, manly voice turning back to an infantile state suggested, first, that vocal ability and independence were equally connected to virility and, second, that the evolution was circular. Old men were, like children, devoid of independence, and this was signaled by the weakness and high pitch of their voices. In the nineteenth century, this was connected not only to the state of their body and the other senses (as was the case in Shakespeare's account) but also to the state of their larynx. In old age, the well-developed organ would lose its strength. Men would go through a second voice break, as part of a form of second puberty announcing the end (instead of the bloom) of their sexuality.[76]

The reference to a second voice break was a powerful one, because the vocal changes boys underwent during puberty were universally described as the most important episode in their vocal development. It was described as a traumatic period, in which boyish pleasures had to be forsaken, and a masculine identity was gained. As Frederick Helmore put it, there came a time in life in which "childish things" were to be put away and "the beautiful boy's voice, like Solomon's enjoyments when no longer attainable, is only 'vanity.'"[77] The metamorphosis was seen as a difficult and often painful one, and many voice teachers were convinced that young male voices had to be rested during the voice break. Continuing to sing during this period, according to some of the teachers polled by Emil Behnke and Lennox Browne, appeared to them as "certainly useless and probably injurious," a practice that would "ruin the finest organs," and that was "the greatest folly imaginable." In answer to the question of whether a boy should sing during puberty, many (though by no means all) seemed to agree with a certain Reverend R. Willoughby Barret: "Certainly not, if he wishes to have any voice as a man."[78]

For young male singers, as these answers already allude to, the break would have appeared as a loss of identity (and possibly a loss of income). Enrico Caruso's *How to Sing*, which explains his singing method in part through anecdote, recounts at some length how the great tenor-to-be mourned the loss of his treble voice, which "plunged him into deepest despair."[79] For girls, puberty was thought to have less impact on the voice. Some authors suggested that this compensated for the greater traumas suffered in other corporeal changes (most notably the onset of the menses), but the general discourse was that women's voices simply did not differ that much from girls'. Their larynxes remained relatively small, their vocal cords short, and the pitch of their voices quite high. According to Frederick Helmore, "The size of the larynx in boys and girls is pretty much the same until the age of fourteen or fifteen, when the change, mentioned in the preceding chapter, takes place. The vocal cords, at that time, lengthen very considerably in the male larynx but very seldom to any great extent in that of the female."[80]

As a result of these observations, it was concluded that women's voices were simply less developed than men's: they could be moving, manipulative even, but did not exude strength or command attention.

This gendering of vocal identity was extrapolated to delineate the finer distinctions between the voice types of singers of both genders as well. Men with a high-pitched voice—especially those using the eerie sound of the falsetto—could easily be labeled effeminate. It was surmised that their larynxes were smaller and their vocal cords shorter than those of basses and baritones. The concept of development was applied within one gender as well: the low male voice was more developed—and therefore more manly—than the high one. Conversely, the high-pitched female voice was not only the least developed, but also the most feminine. Lower female voices were associated with the quasi-masculine sounds that were expected in old age, but also emerged from contraltos, who were described as grave, heavy, and even masculine. Or, as Henry Wood put it in a humorous take on the received wisdom about vocal types in in his *Gentle Art of Singing*, "the real, thick, woolly, hooty, breathy, melancholy-toned, heavy contralto is always threatened with immobility. These singers seem to love stagnation, to be happy in gloom and stodginess."[81]

The only voice type that occasionally escaped the linear image of gendered pitch was the heroic tenor. Tenors were sometimes jokingly referred to as the "third sex," but the power and youthful sound of their voices was highly prized, and this accorded them a measure of virility. Destined to represent "le jeune premier" onstage, the powerful tenor voice signaled both heroic action and heteronormative romance by the end of the nineteenth century.

The reception of the heroic tenor shows that the distinction between voice types rested on more than pitch alone. As Clara de Chatelain put it, "It is the quality and not the extent that should be consulted."[82] Indeed, many teachers of canto described the difficulty of assigning someone to the correct voice type:

> What kind of voice has he or she got? This is generally believed to be the most important question in the beginning of voice-culture, a question that must be answered before teacher and pupil begin their work. Like many pet theories that have become traditions, I cannot accept this as a reliable opinion. I know many cases where teachers have made mistakes as to the classification. This is a serious matter for two reasons: First, because such a mistake is very injurious to the voice; and, secondly, it is usually a long time before the pupil will find out the error.[83]

Leo Kofler was hardly alone in his assertion that diagnosing a young singer's *Fach* was as important for their health as it was musically. Assuming the wrong vocal identity could be harmful to voice and body, and effectively rob a singer of his or her true voice. Moreover, the effects of this loss of vocal identity would go beyond the merely physical, as voice types were thought to be connected to character traits and emotional behavior as well. Consequently, having to voice the wrong roles (whether or not onstage) would be especially harmful. Enrico Caruso warned singers against trying to sing outside of their range or character, asserting that:

> One should never try to change the tessitura or natural character of the voice. A voice will become higher just when it should, by the development due to rational work and never by forcing it. Nothing is easier than to force a voice upward or downward, but to cause it to "recede" as it were, in either direction is another matter. A baritone who tries to increase his upper range by main strength will surely in time lose his best lower notes, and a light tenor who attempts to force out notes lower than his range will never be able to sing legitimate tenor roles, and after two or three years may not be able to sing at all.[84]

Voices, even overtly cultivated singing voices, were thus expected to express personality and emotion. Likewise, specific voices were expected to be accompanied by visible traits pointing to the same personality and social role. Higher voices were connected to youthfulness: sopranos and tenors would usually be cast in the roles of young lovers. Singers possessing these voices were therefore encouraged to cultivate athletic, youthful bodies—but to an extent, it was also assumed that they would already naturally conform to these visual esthetic standards. Because their larynxes were small and youthful, and obviously part of their body, it seemed logical to assume that the rest of their organs and muscles would share the same characteristics. Likewise, basses were expected to be older, heavier, more statuesque in their bearing. The examination notes for song and declamation at the Parisian Conservatoire throughout the nineteenth century are replete with references to the need for singers to have a "bon physique," one matching both the sound of their voice and the role they might play onstage,[85] even though, as Emil Behnke so elegantly showed in his onstage experiments, voice types and even gender could not be distinguished by ear alone. A very elaborate discourse was cultivated to establish a pattern of expectations that would encourage an almost synesthetic perception of voices: hearing a heroic tenor would call the young, broad-chested prince to mind, and seeing an imposing, heavy, bearded man would lead one to expect a bass voice. These cultural stock images were not exactly new, but in the nineteenth century they gained legitimacy through elaborate scientific observations and statistics.

Vocal Propriety and Deviation

Nevertheless, observers did occasionally run into voices and bodies that did not fit the mold, but were also not to be classified as pathologies. The occasional male soprano or female tenor—described with relish by voice therapists and physiologists—put a strain on the apparently seamless logic linking pitch, corporeal characteristics, and personality. These men's and women's voices were recorded as fascinating cases, but authors seemed to hesitate to offer an explanation for their exceptional qualities. They were described as freaks who vocally defied the carefully researched laws of nature, and whose sexual identity was therefore under suspicion as well. In his clinical manual on vocal pathology, James Downie, for example, felt compelled to note that a persistently high-pitched voice emerging from a male throat need not be the result of sexual abnormality, while using the suggestive term "eunuchoid" to describe it, and including the case in a chapter on neurosis: "A persistent high-pitched puerile or falsetto voice in man is occasionally met with. It is not dependent on any detectable physical defect in the larynx, nor is it associated with any sexual abnormality."[86]

Beside the stage, other vocal deviations were connected to personality traits and corporeal imperfections: a host of afflictions that we would now classify as speech impediments were connected in the nineteenth century to an almost holistic understanding of a dysfunctional vocal performance, depending on both corporeal and temperamental characteristics. Definitions of stammering, in particular, underwent an important change around 1860, and its occurrence was connected to a specifically modern identity: that of the middle-class, independent man. Statistics compiled earlier in the century indicated that women, savages, and the lower classes stammered considerably less.[87] This presented researchers with a conundrum: the affliction was generally understood as a result of nervousness and therefore would logically have been connected to femininity (or indeed effeminacy, as some of the harsher stammering cures suggested). However, as the disease seemed to be almost exclusively connected to the professional classes, other explanations were sought—and found—in the second half of the nineteenth century. Stammerers were, according to Beasley, of more than average intelligence. In the (imaginary) interview with the author, it was asked if a quick apprehension or rapid thought has anything to do with stammering.

> *BEASLEY:* Yes. When thoughts come too rapidly, they cannot be spoken in time. Before they are uttered they are forgotten and thus hesitation ensues.[88]

Others, like William Ketley, suggested that the excessive strain of modern life was more keenly felt by the professional classes, and resulted in stammering. Ketley devoted a whole chapter to the stammerer's identity as a "victim of civilization," in which he established that city folk stammered more than countryside dwellers, Northern Europeans stammered more than those in the South, and the English stammered the most of all. All these differences were in some way ascribed to the march of industrial development and civilization:

> Civilisation carries with it in the upbringing of children many other factors predisposing to neurotic affections when regarded in comparison with the lives of children and savages or uncivilised races who are brought up amid surroundings and conditions of perfect freedom. Some philosophic soul has said that "when the monkey blushed man was born." Whether this be true or not, it undoubtedly is true that when man first blushed the stammerer came into being. Blushing, nervous dread, hesitation, are all steps towards stammering, and all are due to the repressive influences of civilisation, aggravated by the wear and tear of modern life, with all its erotic and neurotic tendencies.[89]

Rather than approaching stammering as a deviation, then, these authors suggested that some vocal pathologies were intimately intertwined with one's professional identity. As healthy voices could express a number of personal, emotional, and corporeal qualities, the pathological voice betrayed one's status and social position. The very specific condition of *dysphonia clericorum* is a particularly clear example of this, understood as the result of the speaking patterns of the clergy (or, more widely, those in professions that necessitated bursts of public speech), and thus read as a symbol of those professions.[90] A stammer, paradoxically perhaps, could therefore become a sign of privilege: the typical stammerer was male and middle class and would possess all the desired characteristics of his social group: intelligence, rationality, courage, steadfastness, and self-control. This last quality, especially, he even possessed in greater measure than his peers, as exceptional discipline was necessary to deal with a stammer on a daily basis: the many breathing exercises, gymnastics, and reading examples in manuals demanded commitment. Despite the imperfect and unconventional character of his vocal performances, a stammerer's speech signposted the desired qualities of modern speech in its continuous performance of the overcoming of an exclusively modern, middle-class obstacle: the struggle for control was what expressed the speaker's mental alacrity and his rationality.

The stammerer thus used unconventional means to reach a thoroughly conventional goal: an individualized sound of rationality. If there was something like an ideal voice in the second half of the nineteenth century, that would be it: a vocal sound expressing independence and reason. The voice was capable of an impressive (or indeed alarming) array of emotions. However, not all of these were equally valued, and not all were compatible with the privilege and propriety connected to middle-class masculinity and its connotations of political and social power. As the issue of stammering shows, the vocal expression of rationality and independence had no clearly defined sound—the rational modern man strived to express his individuality as much as his conventionality through his voice. Yet, the range of possibilities was quite clearly circumscribed—and became more so through the continued compilation of statistics, the medicalization of the voice, and the precise delineations between voices connected to different identities.

The continuous tension between individuality and propriety dovetailed neatly with other facets of the modern condition: similarly paradoxical expectations existed for clothing, facial hair, musculature, political discourse (which had to be rational and deeply felt at the same time), and domestic affection (deeply private, yet governed by very strict rules of propriety). In the case of the conventionally individual voice, its paradoxical nature was also connected to the new ways in which the voice was thought to be lodged in the body. In the first half of the century, the corporeality of the voice was projected almost entirely on the larynx, tongue, and lungs. These were treated as a mechanism that—much like a musical instrument—would produce tones according to unchangeable natural laws. The voice was seen as universally human (and indeed as what distinguished humans from animals), and as something that could be observed in every human throat. By the 1860s, this image of the voice started to shift as researchers espoused a more holistic view of the body, of which the voice was a part. In order to control the voice, control over all muscles and organs was suggested. Rather than assuming a universally available mechanism, experts surmised a universal human will to speech. This did not dislodge the voice from the body (or from the so-called laws of nature), but it did connect the act of vocalization to the will and therefore to the speaker's self-discipline and personality.

Rules of propriety and social status provided a level between the universal and the personal, which in the nineteenth century were so intimately intertwined with views on this holistic body that they appeared as natural: modern science endeavored to systematize and quantify the (vocal) difference between men and women, children and adults, laborers and lawyers, and even the finer distinctions between voice types. Identifying as a mezzo-soprano in this period provided one not only with specific roles onstage, but also with a certain set of gender- and age-specific physical and temperamental attributes. Most of these would have been established as cultural tropes earlier in the century, but experimental, statistical, and pedagogical endeavors in the latter half of the century granted them a greater authority and a sense of modernity: they were strongly connected to the modern self and its controllable, cultivated, yet natural body. Although detailed research on the larynx played a role in the naturalization of these distinctions, scientific fervor went far beyond the specific and the anatomical. General body shape, personality traits such as vanity or a hot temper, hair color, and so on, were included in the calculations as well. Any characteristic that could be read and interpreted as part of one's unique identity was seen as relevant to the cultivation and production of the voice.

In order to observe the modern voice and its multiple different properties, all the senses were engaged. The conscious and active experience of vocal sounds—either as a vocalist or as an audience—gained importance in all disciplines concerned with the voice. Specialists relied not only on increasingly sophisticated machinery to observe different aspects of the voice, but also on their own somatic experiences and those of their patients and pupils. As expertise on the voice became more closely related to vocal (somatic) experience, the methodological struggles between elocutionists and physiologists seemed to cool down. Both expertise and experience—as symbols of respectively fundamental and applied science—were increasingly shared between disciplines. Moreover, the lay reader, supposedly a modern, independent vocalist, was increasingly encouraged to deploy his own experience in order to build up his own expertise in vocal production. The will to speech was seen as natural and universal, but exploiting it necessitated conscious engagement with the body, careful control over vocal practice, and specific knowledge. As John Bishop pointed out, conditions that affected the "power of volition" also had deleterious effects on speech.[91] For the civilized, modern self, the road to natural and healthy voice production led through practices of control, conformation, and discipline.

Statistics and simplified schematics encouraged these self-determining individuals to conform to what was considered normal, but did not simply reduce them to generic numbers within a system. Rather, they encouraged readers to contextualize their own experiences in the context of a more collective understanding of their own voice. In conjunction with the heightened attention to personal perceptions and somatic experiences, researchers shifted their focus from the vocal apparatus to the living voice. Experiments were no longer held with dead larynxes, but with living vocalists whose vocal cords were observed through the laryngoscope. It was the production of the voice that came to be observed, then, rather than the organs producing it—which subtly changed the definition of the voice itself. It became a performance, rather than a thing. "It must always be borne in mind," John Bishop reminded his readers, "that we have not to deal with automatic functions, which, once set in healthy action, continue like the movements of a watch; but with mechanism, the movements of which are placed under the control of the voluntary system, and subject to the irregular impulses of the intellectual processes."[92] In the beginning of the nineteenth century, the voice had become corporeal, almost tangible, as its connection to the specific layout of the human throat was established and studied. In the second half of the century, the voice came to be seen as a product, rather than as a part, of the body and its governing matter: the brain.

INTERMEZZO 3

Those Who Have Suffered Know Best

Benjamin Beasley's Stammering Treatment

In 1902, Benjamin Beasley published his *Reminiscences of a Stammerer.* These reminiscences, according to the author's son-in-law,

> show Mr. Beasley as he was; a man of vigorous frame, strong will, iron determination, a man of great intellectual capacity who would have made the business in which he was engaged in early manhood a huge success but for the unfortunate impediment which crippled and handicapped him at every turn, and who, having cured himself, turned to account the discovery he had made and established the greatest school for the cure of stammerers that has ever been known, to the incalculable benefit of thousands.[1]

Beasley's reminiscences dwell, in large part, on the author's unhappy years as a young stammerer, because he felt that the account would prove that stammering caused nervousness, rather than being caused by it. In an earlier prescription of his methods, Beasley wrote: "My own personal experience unquestionably proves to me how fearfully one may be afflicted with the impediment without having reason to attribute it to nervousness. As a child I was very healthy and robust, and at the age of five or six years was remarkable for my fluency of speech and perfect articulation."[2]

Stammering only afflicted him later, and would influence his demeanor—rather than the other way around:

> Although I think I may say I was not at all a bad-tempered lad, I was sometimes driven into uncontrollable anger when chaffed about my infirmity; indeed, when at school I on several occasions fought with bigger and older lads than myself, in consequence of their allusion or imitation of my defect.
>
> Lads are prone to make game of an infirmity out of sheer thoughtlessness, but on the whole I did not suffer much on that account while at school, as, fortunately, I was very robust and much stronger than the generality of boys of my own age, and this no doubt in a measure lightened the burden it would otherwise have been.[3]

The author's "stammering years" are described as a long period of misery, alleviated only by one teacher, "a splendid old fellow," who awakened young Beasley's taste for reading, but did not cure his speech impediment. Even in business, as a young man, Beasley indicated that he would "generally find a way out of it by getting someone else to talk for me."[4]

Beasley noted that his parents had done "all they could to get me cured, but unfortunately many of their plans were not only unsuccessful but injurious. I was sent to different schools where the masters had an idea they could cure me, but in several of these trials I was made worse."[5] He ultimately endeavored to devise his own methods to deal with his predicament, working his way through the wisdom and science of the day:

> I found in books various speculations and theories, some of them very learned and scientific, but few of them practical. They taught me nothing, and only told me in scientific method and terms that which I already knew. I gained little or no help from books.
>
> My next experiment was to observe carefully those who spoke well, and note if I could see the difference between them and myself. If there were any well-known speakers or readers to be heard, I made it a rule to go and hear them. I endeavoured to carry in my mind anything that struck me which might be of help to me. I practised speaking and reading before a looking-glass, endeavouring to find out the difference which caused them to speak with freedom and myself otherwise, but I found no aid in that. . . .

> Among my intimate acquaintances were two actors, who I think were the finest readers and speakers I ever knew; they belonged to the old school of Shakespearean actors, and in their day were among the celebrated. I never talked with them without feeling their influence on my own speech—their full, round, deep voices, their perfect articulation, and their musical inflections always left an impression on me. I used to practise reading as much in imitation of them as possible, and I always fancied that I spoke better after such drill, but although I could read fairly, almost perfectly when alone, I could not do so in the presence of others.[6]

From his dilettantish study, Beasley claimed to have devised a cure that was beneficial not only to himself, but to others as well. He

> treated most successfully cases which have baffled all previous attempts to cure, some of them having succumbed to my treatment in a very short time. I can only attribute this unexampled success to the fact of *having been a most inveterate stammerer myself,* and knowing, and thoroughly feeling, the sensations of a stammerer. This coupled with the fact of *having cured myself,* is a sufficient guarantee of my capability to successfully treat others similarly afflicted.[7]

His reminiscences made at least some of his methods available, although Beasley indicated that he was "convinced it is impossible to impart my system by written instructions, *in a manner thoroughly satisfactory to myself, or so as to afford sufferers the utmost benefit.*"[8] Some of his instructions were presented as an imaginary conversation with the author, in which the reader could imagine he was being personally instructed:

> *I write this Chapter in a conversational form. The questions are similar to those frequently asked of me. My replies will give the necessary information to correspondents, and thus obviate the necessity of writing long letters.*
>
> Can stammering be cured?
>
> It can: for many years I held a different opinion.
>
> Why so?
>
> Because for many years I was an inveterate stammerer myself, and now I do not stammer.
>
> Did you cure yourself?

Certainly.
How did you do so?
By first learning the cause, and then observing others who spoke well.
What is the cause of stammering?
The causes are many, but may be summed up as endeavouring to speak in an impossible manner.[9]

Far better than this imaginary conversation with the author, however, would be a real stay at his stammering institution in Brondesbury, where his "perfected system" was taught by his daughter, her husband, and their children. The institution, Tarrangower, not only provided a place to adopt the principles of the Beasley method, but also reflected the image of Beasley's younger years. It presented itself as a place where a "robust lad" would feel at home and could develop himself. Its wide grounds and airy rooms were commensurate with current ideas on health and hygiene and supported a self-representation of Beasley and other stammerers as epitomes of nature and health.

Although Beasley's followers seemed to be influenced by the late nineteenth-century wave of hygienism, the author did not specifically align himself with any theory or science, medical or otherwise. He prefaced his publication of his stammering treatment with: "Believing, as I firmly do, that there is not the slightest reason for supposing it to be a question either for the physician or surgeon, I have not, in treating the subject, chosen to express myself in medical or other scientific methods and terms, but have endeavoured to clothe my opinions in so plain a manner that I may be comprehended by all readers."[10]

He was particularly keen to distance himself from various "speculators on stammering":

> During the past three hundred years, speculators on stammering have been very numerous. Many of these held opinions quite as absurd as some of the ancients, both as to its cause and treatment. It would take up too much time, and serve no good purpose, to consider the various opinions on the subject, which are to be met within the works of Lord Bacon, Joseph Frank, Dr. Lichtnger, Vosin, Hervez de Chegouin, Dr. Schulthess, Dr. M'Cormac, Professor Müller, Dr. Marshall Hall, Serres, and many others. It is, however, somewhat amusing, and at the same time shows what ignorance of the real cause of stammering has prevailed, to know that such remedies as flogging, shaking, rubbing the tongue with salt, charms, and others quite as ridiculous, are recommended.[11]

Figure 7. Benjamin Beasley, *Reminiscences of a Stammerer*, London: Roxburghe Press, 1902, frontispiece. Courtesy of the Wellcome Collection.

Claiming his own method to be uniquely natural and uniquely effective (a discourse he shared with most of his colleagues), he presented himself as the therapist who succeeded where others had failed, describing cases of desperate students who would finally find their cure in his methods. One after the other, his pupils were described as having shared the author's traumatic experience of different harmful therapies: "Another pupil who had been forced to give up his situation as managing clerk in a large establishment in a provincial town, applied to me. He told me he had been more than twelve months in London and elsewhere, seeking a cure for his impediment. He had seen many professors, and had purchased every book he could hear of which treated on the subject. His experiences were rather varied."[12] Beasley derived his authority (or at least his confidence) from his own experiences as a stammerer. Being a fellow sufferer not only provided him with an authentic insight into the nature of the affliction, it also signaled his superior intelligence. According to Beasley, "As a rule, stammerers are more intelligent than those not so afflicted. This is easily accounted for, as it is natural that those who are, as it were, in a measure cut off from society, will be more likely to cultivate the habit of thinking, and thus strengthen their minds while finding a solace in reading and contemplation, than those not so debarred from communion with their fellow-creatures."[13] Claiming extraordinary intelligence for stammerers—combined with the assertion that the stammerer could be robust and strong—was perhaps the most distinctive and effective part of the Beasley method. It aimed to change his pupils' perception of themselves, and, at least according to later therapists dabbling in psychology, such as Alfred Appelt, this would lift the curse of speech phobia and allow them to speak freely.

Beasley would go on to identify as a stammerer throughout his life, most notably during his numerous public lectures. He encouraged pupils at his institution to do the same on publicly held "speech days." These days of public speaking, according to a reporter from the *Birmingham Daily Gazette*, demonstrated "the success of Mr. Beasley's system . . . beyond a doubt . . . and as the relief afforded is not temporary, but permanent, many ladies and gentlemen have occasion to remember with gratitude the beneficent course of treatment to which they were subjected at the hands of Mr. Beasley."[14] This, then, is what the stammerer at Tarrangower looked and sounded like, according to its director: not a frail frame turning into a nervous wreck, but an imposing man speaking freely.

CHAPTER 4

Thinking Voice

Civilization, Vocal Expression, and the Brain

Throughout the first half of the nineteenth century, the location of the voice in the body had been a subject of research and discussion. By the 1860s, the frantic search for a seat of the voice and the burning desire to see the larynx at work had largely been fulfilled. The general consensus was that the voice was produced by the body in its entirety. Physiologists studied not only the different organs involved in voice production but also the influence of the muscles on their workings. Physicists noted that reverberation played its part in the sound and quality of the voice. This involved the whole body, and even the matter surrounding it, in processes of voice production. Meanwhile, less tangible qualities, such as courage, nervousness, or passion, were thought to have their influence on individuals' voices as well. But in the second half of the nineteenth century, this balance between sites of knowledge about vocal production would change again. If the period between approximately 1790 and 1860 represents the rise of vocal physicality and materiality, culminating in the invention of the laryngoscope, then the latter half of nineteenth century, inaugurated by Paul Broca's paper on his discovery of the speech center in 1861, represents a radical new way to locate vocal management in the brain. Broca's discovery did not prompt immediate upheaval: the different fields of knowledge concerning voice production and health were too well-established by 1860 to be thrown into disarray so easily. But the notion that it was the brain that initiated and shaped vocal utterances caught on in both expert and amateur circles. By the beginning of the twentieth century, even the most eccentric authors would reproduce the increasingly popular understanding of the brain as a central telephone station.

In 1904, Emil Sutro—a German immigrant in Philadelphia with a keen mind for business and an even keener but quite a bit less informed interest in the science of speech—self-published his study, *Duality of Thought and Language*. Sutro had not been able to find a publisher for his rather outlandish theories on voice and speech, but his dilettantish attempt to come to terms with the scientific understandings of voice production at the time are a particularly florid representation of the shift that had taken place in this field in the second half of the nineteenth century. In the decades before his work, according to Sutro, "The voice has been a desolate creature living in a hut, which was filled with the foul exhaled air from the lungs, and in which it led a precarious existence, subsisting on such things as mucous membranes, muscles, sinews and the like. An abject creature, a pitiful existence."[1] In his own studies, Sutro was determined to look beyond the "hut" and approach the voice as emanating from the brain, which he considered "the central station, rather, as it were, of a net of telegraph or telephone wires, in which messages are received and from which messages are sent out in a thousand directions."[2] Sutro's liberation of the voice from its bodily confinement, toward a freedom to roam at the speed of electricity along wire-like, telephonic nerves, illustrates how notions of modern science were shifting. The messy world of organs on the operating table and patients carrying their diseases to doctors' offices would of course continue to exist, but science would increasingly become a matter of detached laboratory work.[3] Clumsy statistics with limited numbers of patients would be replaced by much more impressive tables of isolated bacteria and proteins.[4] Nevertheless, shifting the seat of the voice from the lungs and throat to the brain only partially disembodied it. The brain, after all, was a material thing as well, and undeniably part of the body. Sutro's further specification, of a brain that is not a repository of information but rather a center for exchange, took the disembodiment of vocal practice even further. The voice, in this model, is not only not seated in the lungs—it simply has no seat.

Sutro's notion of the immateriality of the voice was taking a stand on where research on the voice belonged, and which scientific fields were to dominate discussions on the nature of the human voice. Despite its eccentric tone and implausible theories, Sutro's *Duality of Thought and Language* reflects many of the tensions that characterized the debates surrounding vocal health and propriety in the latter half of the nineteenth century—and the tensions that will be at the heart of this chapter. Focusing on speech, I will first describe when and how sciences of the mind (neurology and psychology) rose to prominence in discussions on the voice and how Broca's discovery reshaped the scientific fields in which research on the voice was undertaken. The discovery of a speech center in the brain led researchers to pay more attention to the connection between language and voice (rather than, say, fluency or pronunciation, although interest in those fields remained strong throughout the century). The second half of the chapter therefore focuses on practices of distinction and expression of identity through speech and linguistically particular instances of vocal management.

Seated in the Brain

As Sutro's ambiguous pronouncements on the brain and the mind show, the shift from an organic/physical understanding of the voice toward a renewal of the disembodied one was neither simple nor linear. Moreover, innovative perspectives on voice and speech did not necessarily appear as particularly new, as they could seem to hark back to older theories on sound and voice production. The notion of electrical energy carrying information took time to develop as a plausible explanation for aspects of human life, and easily slipped into older discourses of voices being carried on breath.[5] The process of change in understanding voice and sound was at least partly accompanied and inspired by numerous technological inventions that promised—and eventually managed—to produce apparently human voices from a machine, even if human bodies would generally still have had to generate the original sound for machines to record and replay. In 1860, Édouard-Léon Scott de Martinville made what is now known as the first machine recording of a human voice. Like earlier, human-generated recording methods, his phonautograph transcribed sound but could not play it back. But more elaborate recording machines were invented soon after, and put to work preserving and storing famous and remarkable voices.[6]

These technological and scientific changes pulled the human voice in two directions. On the one hand, the regained stress on the immaterial, nonbodily character of the voice brought back associations with eerie, (quasi-)religious phenomena. The notion that phonographs could somehow preserve the voices of the dead was thoroughly Victorian,[7] but also referred back to early modern notions of the voice as a manifestation of the soul. On the other hand, machines' increasing ability to produce human speech allowed for an increasingly detached or disinterested treatment of a phenomenon that was, perhaps, not quite so intrinsically human after all. Lodging the voice in machinery, or in the brain, rather than in the voice box or the lungs, strengthened the ties between vocal sound and human individuality or personality, by connecting the voice to processes of thought rather than the organs of speech. However, it also loosened the link between voice and body, which made tethering vocality to humanity more precarious. The two scientific disciplines that would come to claim large parts of the domain of vocal science both struggled to come to terms with these tensions in different ways. Elocutionists and surgeons remained active throughout the nineteenth century, and these fields continued to develop, often drawing on new research regarding the human brain. But it was the work of psychologists and neurologists that would garner more attention as carriers of innovation and change. One enduring result of the rising influence of psychology on the study of speech, for example, was that elocutionists would slowly disappear. One could argue, perhaps, that the profession was rebranded as speech therapy, practitioners of which often continued to have some background in performance in addition to their interests in medicine and therapy well into the twentieth century.[8]

Although the rise of two separate branches of science concerned with both the human brain and its involvement in speech occurred roughly at the same time, it did not produce quite the same tensions as those that had existed between surgeons and elocutionists in the first half of the nineteenth century. Nevertheless, both psychology and neurology had a particular interest in voice pathology and would lay claim to the ability to cure speech impediments such as aphasia, albeit by thoroughly different means. It was perhaps because the psychological approach could so easily be embedded in older notions of an educational/therapeutic approach to speech impediments that its appearance in print, and in the field of knowledge, seemed quite unproblematic. To elocutionists in the latter half of the nineteenth century, the theories and practices of stammering therapists drawing on Jung's theories of association must have appeared as very similar to those of any other colleague who included exercises with word repetition.[9] Neurology, with its exciting and often invasive experiments on a part of the body that had largely remained hidden before, appeared to be a completely new branch of science and therefore unready for application on a wider scale.

Overall, the brain made appearances across genres in the second half of the nineteenth century—as is the case, for example, in this 1852 pronunciation guide for children, in which it was included in a poem about speech:

> You think a fool can live without brain?
> What a mistake; it's not the brain that just won't do
> The problem is in . . . the lungs, no? Now isn't that new.[10]

The notion that speech—and speech impediments—were located or originated in the brain rather than the organs of speech, preceded Broca's discovery. The first half of the nineteenth century saw numerous embryonic/tentative attempts to chart the effect of whatever happened in the brain (and physicians were decidedly unsure about what that was) on the organs and the muscles. Without the metaphors that would come into currency later, in which the brain was imagined as a telegraph or telephone station, or an apparatus sending out electric signals, imagining the functions of the brain and representing them to a reading audience proved difficult. Nevertheless, a number of authors writing on voice and speech agreed that the first spark of speech occurred in the brain. This could be imagined using older vocabularies, such as those of the spirit. According to a French stammering expert in 1847, for example, "The spirit [*l'esprit*] is the whole of all the faculties that have their seat in the brain, such as imagination, intelligence, memory, etc. Therefore, our ideas either emanate from ourselves, or they could come from elsewhere, to the spirit."[11]

The term *esprit*, with its potential spiritual connotations, is used here, but what is referred to is a conglomerate of faculties, closely related to reason: the author does not mention emotion or faith, for example, thus clearly attributing this spirit to the modern, thinking individual. *Esprit*, as it is used here, could perhaps more accurately be rendered as mind, a term English-speaking authors were using at the time as well. They, too, referred to the brain as a place where ideas come alive. If things could "come to mind," ideas could also be "received," either arising from immediate sensations or originating in the brain in an abstract manner.

Once ideas were thus received and "arranged into proper succession"[12] by the brain, they could act upon the body. Much like the telegraphic metaphor still to come, authors imagined the brain at the center of a great network of nerves that connected it to all organs and muscles—or indeed a network of "routes," as one French author on oratory put it in 1841: "I think, messrs. that you would easily agree with me that the voice is to the ear what writing is to the eye; in other words that they both, albeit by different routes, transmit thoughts to the brain."[13] This notion of transmitting thought through language was central to oratorical training, which was often presented as a practice of clothing thoughts with voice. It was also of particular interest to those studying stammering, which they tended to see as a nervous affliction. The designation implied some sort of nervous disposition in stammerers in a general sense, but could also refer much more precisely to the effect of the brain—and particularly an overexcited brain—on the nerves. The intricate web of brain, nerves, organs, and muscles gave a more material—and more convincingly medical—basis to elocutionists' understandings of speech and stammering. These representations of the brain as the seat of intelligence, a receiver of ideas sending signals to the rest of the body, seem intimately connected to further developments in scientific inquiry in which the brain's role as the seat of speech was to be discovered.

The notion of such a seat of speech or a speech center preceded its eventual discovery. In 1841, twenty years before Broca could pinpoint its place in the frontal lobe, stammering professor Claude Vernet identified the brain as *le siège* where stammering took place. More precisely, the essential cause of stammering, unknown at the time, was sure to be found in "the part of the brain that presides over speech," much like the causes of congestion would be found in "the portions of the brain that regulate the locomotive muscles."[14] Although the notion of a speech center was thus gaining traction in the first half of the nineteenth century, its exact position and—more importantly—its precise role in speech production and speech pathology remained unclear. For Vernet, it represented almost unlimited possibilities for finally fully understanding the cause of stammering, an issue that would remain contentious throughout the nineteenth century and up to this day (neurological discoveries notwithstanding). Two decades later, another French medical doctor wrote with similar confidence and somewhat more precision about the "seat of intelligence" (*siège de la pensée*) in the brain, which he attempted to connect to speech: "It was not unknown in cerebral physiology that the seat of thought was located in the upper and anterior part of the brain; this was in fact the only clear and certain fact amidst doubt and obscurity which makes the psychological physiology of the brain somewhat less problematic as a science."[15]

This, according to the author, was all common knowledge. His own contribution to the debate was an exciting one: he would show *how* the ideas generated in the brain were transformed into speech: "The goal of my work is to show that there is, in the anterior half of the brain, an organ of internal speech."[16] This internal speech was to be distinguished from speech as conventionally defined: "the speech of the mouth makes itself heard to others, whilst the speech of the brain can only be perceived by the individual that produces it."[17] The connection between this *parole interne* and intelligence was clear and would be reflected in the proximity of both in the human brain, "in the upper and anterior part of the brain."[18]

These early pre-neurological and pre-psychological studies pointing to the centrality of the brain prefigure the ambiguity that would result from a new perspective on voice production that would only really come into its own from the 1860s onward. As I have argued in preceding chapters, the nineteenth century presents a somewhat anomalous period in thinking about the human voice. It is a period of great attention to vocal materiality and embodiment that had not really existed before modern science and would largely disappear again once acoustic recording technologies became widely available. The first attempts to understand the role of the brain in voice production were swept along in the momentum of discoveries, experiments, and new cures for the physical voice as it was understood in that period. Despite embryonic attempts to represent the brain as a terminal (prefiguring the language of connections and synapses), these early attempts very much saw the brain as a location of voice production with a specific function, analogous with other organs of speech such as the larynx or the lungs. Rather than a radical shift, this move toward the brain could also be understood as part of the slow process in which physicians came to think of voice production as involving the whole body, and not just one isolated box.

Most indicative of the ambiguity of the evolution of sciences surrounding the human voice is perhaps the timing of the two most important scientific discoveries in the field. The laryngoscope, which almost figures as a symbol of the embodied, material approach to the voice after a long period of assigning spirituality or soul to the voice, was presented to the world in the mid-1850s. It knew almost immediate popularity—manuals for its use were published, polemics over its real inventor surfaced and resurfaced, and the instrument was used by physicians, therapists, and singing teachers across Europe and the United States. Its influence was immediate and remained visible throughout the century. The other great innovation in the field, the discovery of a speech center in the brain, was quite different in nature and slower to catch on. Although Broca's discovery seems to belong to an entirely different way of thinking about vocalization, and therefore to a different era, it was published less than a decade after the invention of the laryngoscope was made public. The two inventions are rarely discussed together, because they are part of such different worlds. Neurology/neurolinguistics and laryngology developed into separate disciplines in the twentieth century, with different methods, approaches, and largely different goals. In the context of this book, however, and in the context of the nineteenth-century field of study of the voice, in which disciplines were not fully formed and were harder to distinguish from one another, they do relate to each other and thus belong to the same history.

So, what was it that Broca discovered? His findings were presented to the Société Anatomique de Paris and the Société d'anthropologie in the early 1860s. There, in lectures to his peers, Broca explained that "we" (i.e., humans) are "left-brained," at least when it comes to speech as well as "much more simple and more crude acts."[19] Not only was speech directed by the left hemisphere, but so was "writing, drawing, embroidery, etc."[20] This location of a speech center in the brain's left hemisphere, according to Broca, was not a simple matter of anatomy—both parts of the brain are anatomically identical—but because "the earlier development of the left hemisphere predisposes us, in our first fumbling attempts, to perform the most complicated material and intellectual acts with this half of the brain, among which we must certainly include the expression of ideas by means of language and, more specifically, of articulate language."[21] Aware that this notion of localizable parts of the brain with different functions was likely to encounter some resistance, Broca stressed the experimental, empirical aspects of science: "Observation is of course superior to theory, and sometimes we have to bow before a fact, no matter how inexplicable or paradoxical it may seem to us."[22] His observations had taken place through the autopsies of individuals who had suffered from aphasia (*aphémie*) in life, in which he had found lesions in particular areas of the brain. Much like the vocal physiologists had done before him, he focused on pathology and abnormality—which could be found and then observed—in order to come to an understanding of normal speech or the ways in which brains deemed normal could produce it.

The actual observation of a speech center in the human brain is now indelibly connected to Paul Broca and his presentation to the Société d'anthropologie, which, although it was not exactly widely publicized initially, certainly made him an authority in the burgeoning field of (linguistic) neurology.[23] At the time, however, Broca's claim to the discovery did not go uncontested. As had been the case for the laryngoscope, the invention of the speech center cannot be quite so simply assigned to one author or scientist. As Heini Hakosalo paints vividly in her history of the study of aphasia, *On Speaking Terms*, "boundary work" and moments and practices of contestation and conflict between scientific practitioners and their discoveries were central to the field. Broca himself addressed questions of primacy in his 1865 lecture in Paris. "I do not enjoy discussions of primacy" he stated, after which he nevertheless set out to claim his discovery as an original one.[24] At issue was the work of Marc Dax on lesions in the left hemisphere, which had been published very recently (in 1865) by his son.[25] "Dax fils" had made great efforts to ensure his father got the recognition he thought he deserved, pushing Broca into a position of self-defense. In his lecture he stated he had not "sinned by ignorance or willful omission": the older Dax's work had simply been unknown and therefore impossible to be aware of or referred to. The question of the originality of his discovery was not just a matter of primacy though. Broca's particular methods of observation and his explanation of a speech center also presented a departure from other, older theories on particular parts of the brain representing particular functions. Earlier attempts to pinpoint the place of speech in the brain that had been connected to phrenology and neurological research initially had a hard time shaking off connotations with the theory—which had fallen into disrepute by then. As German pathologist Adolf Kussmaul recalled in his 1877 book *Speech Disorders*, the reception of the eventual publication of Dax's work was not unproblematic: "The reporter Lélut wasted but few rude dismissive words and declared every attempt to localize language to be mere phrenology."[26]

The contested history of the repeated attempts of natural philosophers and nineteenth-century physicians to locate articulated language in the brain was laid out by Dr. Albert Carrier in 1867. His *Study of the Localization of the Faculty of Speech in the Brain* (*Étude sur la localization dans le cerveau de la faculté du langage articulé*) was published at a time when, according to the author in the introduction, the number of observations would soon turn theory into doctrine.[27] His account of the different steps leading up to Broca's "definitive" work supported Broca's narrative of scientific legitimacy. The work had also benefited from "the wise and well-meaning advice" of Paul Broca, which rather points to where Carrier's loyalties lay. Nevertheless, Carrier's study is not a hagiography of Broca—it paints a reasonably broad picture of the recent and less recent history of "organology," "the principle of cerebral localization." The medical interest in the brain as a seat of knowledge and language is traced all the way back to the ancient Greeks: "Hippocrates already" had made observations based on the notion that "the brain is the organ of thought."[28] Since those early beginnings, the ambition to precisely locate intelligence in the brain had waxed and waned: "seen as utopic by some, and a hallucination by others, it had never been rigorously developed and had fallen into oblivion."[29] It was only at the beginning of the nineteenth century that the project was taken up by "a man of genius" who would kick-start the suddenly much more fast-paced chain of observations and discoveries that would take place across Europe, but particularly in France. In Carrier's account, the first "genius" who "stated this principle of localization with all the precision and importance it deserved,"[30] Franz Gall, was followed by a handful of other great men who consecutively worked toward a common goal. Although Lélut appears as a somewhat critical voice, the sequence can hardly be read as debate. It is a narrative of quasi-unimpeded progress, from "utopia" to "doctrine."

In reality, of course, things did not go quite so smoothly. As noted above, Broca failed to take note of Dax's previous work (for whatever reason), and the notion of primacy mattered, if perhaps not to anyone else, to Dax fils and Broca himself. Other authors on the subject seem to at least have been aware of this tension, choosing to refer to both in one breath when referencing the discovery of the speech center—possibly to avoid having to take a stand on the issue. And while fellow specialists followed Broca's discoveries and turf wars with interest, his speech center did not exactly take the rest of the (vocal) world by storm. Carrier's characterization of the field of research as a largely French affair seems to hold after 1863 as well. It is only from the 1880s onward that Broca starts to show up with any frequency in English publications on vocal science. Leading German physician Adolf Kussmaul's brief history of the discovery of the speech center in his book on speech impediments,[31] for a German audience, stressed the importance of a small group of experts at the root of that discovery even more.

Although the terminology of the brain as the seat of intelligence and language was widespread, the connection between such ideas and a concrete neurological approach to language, or intimate knowledge of the findings of one of its heroes, is far from clear. Nevertheless, Broca's ideas and observations did impress his peers and would eventually gain traction outside the small circle of specialists. In 1892, for example, Broca's work was cited quite extensively by the French retired physician Abraham Netter, the author of *Internal Dialogue and the Soul* (*La parole intérieur et l'âme*), whose liberal interpretations and reiterations of the theory show to what extent Broca and his speech center had become the object of common parlance. Broca's terms would thus gain none-too-fixed meanings that would lead to associations with highly valued scientific work for an audience that had no specialist knowledge of neurology, but rather an appetite for knowledge about the brain (which they likely knew was at the frontier of scientific research). According to Netter, Broca's discovery also had a "spiritualist application," for example.[32]

The discovery and subsequent spread of the notion of a speech center in the brain was, all in all, not a general revolution. As noted above, Broca's findings initially did not spread beyond a small group of experts, and even later, the speech center became a vague notion of language formation to be obliquely referred to by most experts on vocalization and articulation (rather than a phenomenon they thoroughly understood). The speech center was, of course, mainly at issue for those whose work was situated at the meeting point between voice and language, rather than, for example, those concerned with esthetics or timbre. Or, as Frederick Mott summarized the issue in 1910:

> The conditions necessary to the existence of speech arose with articulation, and it is intelligence that has converted the vocal instrument into the speaking instrument. For whereas correct intonation depends upon the innate musical ear, which is able to control and regulate the tensions of the minute muscles acting upon the vocal cords, it is intelligence which alters and changes the form of the resonator by means of movement of the lips, tongue, and jaw in the production of articulate speech. The simple musical instrument in the production of phonation is bilaterally represented in the brain, but as a speaking instrument it is unilaterally represented in right-handed individuals in the left hemisphere and in left-handed individuals in the right hemisphere.[33]

The original research on the speech center had been concerned mainly with conditions like aphasia—and the notion that the brain was at fault in various speech impediments would gain traction throughout the second half of the nineteenth century.

But the turn toward the brain also indicated a more general shift in how practices of vocalization were imagined. As shown in other chapters, the material understanding of the voice had led to a focus on learning how to use the muscles and the organs involved. This implied that the health, size, and shape of those muscles and organs had a direct effect on the particular sound of the voice (leading, for example, to substantial differences between voices in bodies of different genders). The learning process also meant that the voice would change over time—particularly between childhood and adulthood. Although the performativity of a voice thus embodied was rarely discussed in a direct sense—vocal health as well as vocal difference was generally represented as rooted in anatomy or biology—it was nevertheless clear that certain actions were required to respect and sustain the natural order of things. The discovery of the role of the brain in vocalization in general and speech in particular added a dimension to these demands on the care of the modern individual's voice. If vocalization was (partly) initiated in the brain, that meant that a good voice necessitated not only a healthy body, but a healthy mind as well—a thought that would become particularly pervasive in the diagnosis and cure of speech impediments. These were no longer seen as nervous afflictions akin to chorea, but rather as connected to emotive and cognitive processes (or glitches) of the brain. As Marjorie Lorch has shown in her study of the work of British neurologist Henry Charlton Bastian, diagnoses of faulty language processing in this period were based on psychological models of processing that were believed to be materially rooted in the brain.[34]

The Sounds of Evolution

While, earlier in the nineteenth century, various vocal experiences and practices were believed to inform and influence overall health and to express the speaker's or singer's character, in the second half of the century this gradually started to make way for an understanding of vocalization as the result of various cerebral processes that demonstrated humanity's superior development and evolution. Yet, it was only by the end of the century that the specifically neurological approach to speech really started to spread. It did so partly by the sheer force of Broca's fame—his name would appear in studies by authors whose understanding of neurology seems to have been tenuous at best—but also through the work of authors whose interest in the mind and brain went beyond neurology. In a "clinical and psychological study of aphasia," for example, American neurologist Joseph Collins referred to Broca with more than a passing footnote. However, Collins also claimed a much larger field of knowledge for the study of aphasia, attributing it to "the study of those disorders of speech."[35] Collins's account is shortsighted as a history of the field—which outside of the new and fashionable discipline of neurology had been concerned with far more than speech impediments—but at the same time it takes a broad look at the "acquisition of knowledge" on the faculty of speech by his contemporaries. The inclusion of psychologists in this endeavor is the most striking and novel, as this was a new discipline at the time and its connection to speech and language did not become immediately apparent. For this author, however, the matter was quite clear: "An introductory analysis of the so-called faculty of speech, from the psychological point of view, must consider: (1) the function of vocal articulation and of writing as phenomena of bodily movement; and (2) the sensational, ideational, and other antecedents or stimuli provocative of these exhibitions of complex motorial co-ordination; and (3) the development of the function of speech."[36] All these functions were understood to be developed in the brain but could not be studied by mere observation "of" the brain, which made them the terrain of psychology. Or, as another of his colleagues expressed it, speech and song depended on both a "cerebral" and "psychic" mechanism.[37]

At the heart of the endeavors of these psychologists of speech was a study of the *development* of the faculty of speech in humans or in the human brain. And because the brain itself could give but few clues as to how its newly discovered workings came into being, and how they had turned into intersubjective means for communication, human behavior (including behaviors historical and linguistic) came under scrutiny. Psychology, as a very new discipline, was therefore claiming its place not only in the context of medicine and neurology, but also as a source of knowledge surpassing what had been studied by historians and philologists. According to Albert Lemoine, for example "the problem of the origin of language" was "insoluble as a historical problem" and therefore had to be transformed into "a psychological question."[38] While the study of language and history through a psychological exploration of "human nature" appears as a somewhat overambitious project, its reliance on comparative biology made it a plausible ambition for its time. Moreover, the notion that "the origin of language," like the origin of man, played out at species level and was a matter of human nature may have been an understanding mainly shared among experts. The connection between speech and personality was made on a broad scale, which legitimized psychological approaches to speech. According to physiologist Albert Deleschamps, for example, each word was a "mixed product," psychological and physiological at the same time, that was "made out of the indissoluble alliance of sound and meaning, a phonetic and a psychic element, thought and expression, signified and signal."[39] By far the most common application of psychology to the field of the study of language and speech was in speech pathology—and particularly in the study of stuttering. Much like self-help and therapy gurus such as Benjamin Beasley, those with a psychological approach built their explanations on the premise that the stutterer did not have a weak mind, but an exceptional one. In France, where physiologists like Itard and Colombat had been at the forefront of stuttering research before, Arthur Chervin, who ran his own "stammerers' institute" (l'Institut des bègues de Paris) and was president of the French Anthropological Society, became one of the leading voices in the discipline, and he turned toward psychology for explanations.[40] On the basis of numerous cases in which the fear of stammering produced a more full-blown pathology, Chervin formulated a theory of stuttering as the result of phobia, a condition gradually created out of singular instances like the one of the army officer described below:

> I will cite the case of an officer suffering from a relatively slight stammer who one day, while he was having his soldiers perform marching exercises, told himself: "What if, by chance, I cannot stop them!" The

> soldiers walked in step, and the officer was still preoccupied with the idea that he might, perhaps, not be able to shout Halt! This thought obsessed him. When he wanted to take command, he could not . . . and the soldiers, who were training in a field bordered by a ditch, were still advancing. The officer could clearly see them moving towards and getting closer to this ditch, but he felt absolutely incapable of shouting: Halt! Finally, seeing them on the edge of the ditch, he made a violent effort and uttered a cry of some sort. As the soldiers asked nothing better than to stop, they understood and stopped to the great satisfaction of their unfortunate officer.[41]

Psychology not only offered new explanations for speech pathologies, it provided new solutions and cures as well. Where previous attempts to cure stammering through surgery or didactic methods had failed, it was believed that the new insights provided by followers of Drs. Freud and Jung promised new hope. London-based author Alfred Appelt, who claimed to have "languished in the gloomy and mysterious fetters of this agonizing complaint during youth and early manhood," dismissed most of the techniques of early therapists and stammering professors who focused on practice, and turned toward the seemingly more innovative insights of psychology:

> during the last decade those who have sought a cure for stammering have gradually been led into the right path and have started from a correct basis—namely, that a stammerer is not, and never has been, lacking in the faculty of carrying out all the "movements of speech" correctly. What does happen is that he experiences a momentary hindrance by hampering mental impressions. Thus, intelligent investigators reject more and more conclusively all tiresome exercises which aim at a systematic training of the organs of speech, having at last begun to realize that the stammerer can speak, provided that no psychic influences interfere. Speech drill of any description is seen by such investigators to be utterly superfluous.[42]

All previous methods of reeducation, therapy, or medical intervention could be done away with; all that needed to be tackled was "the problem of dread."[43] Stammering children were henceforth to be diagnosed by means of (psycho) analysis—and specifically, analyses of their fearful nightmares. Curing them would then be a matter of removing "dread" from the associations that were believed to be connected in a speaker's mind with certain words.[44]

Psychological approaches to language and speech had drawn on Darwinian notions of the origin of man to come to new understandings of the origins of language. In such an evolutionary model of human language, articulated speech appeared as only one means of human communication. Or rather, speech appeared as a pinnacle of human achievement in communication—a practice that in its broadest, or less sophisticated, sense was shared by animals. Other ways of communicating fear, anger, or lust (through cries, for example) had been extensively observed in animals and were interpreted as an evolutionary precursor to language. According to, for example, Richard Garner, whose focus was on the speech of monkeys, "the sounds which express crying, sighing, or laughter, may indeed be a faint suggestion of speech, since we infer from them the state of the mind attending the emotions which produce them, yet they are not truly speech."[45]

These ruminations on the communicative quality of speech and its connection to emotion were not overly precise in their treatment of animals but did roughly run parallel with the findings of comparative biologists, who were increasingly interested in animals' ability not only to imitate human sounds, but also to judiciously use vocal sounds to confer meaning. Nottidge Macnamara, for example, described numerous examples of "vocal sign-making" in birds—the ability to associate sounds with objects.[46] In the "origin of language" along psychological lines, however, there were no reflections on animals' anatomical characteristics and their consequences for the ability to speak. The interest of these authors, unlike those of comparative biologists, exclusively concerned the intellectual capacity to speak—and the psychological predisposition to want or need to develop such intricate ways to communicate.

Evolution between different species was, of course, quite difficult to research, as the process could not be readily observed. But, as in other branches of evolutionary thinking, a solution was found by observing the much faster process of human growth. Drawing a parallel between human children's acquisition of language and the evolutionary process of developing language as a species, attention was geared toward babies' hungry cries and their subsequent learning curve toward the expression of different emotions, ideas, and—eventually—their grasp of language. According to Frederick Mott, the process was quite simple: "articulate language is the result of an elaboration in the long procession of ages in which there occurred three stages—the cry, vocalisation, and articulation."[47] Drawing such parallels between growth and evolution was a precarious exercise, as locating the origin of language on the evolutionary ladder as a naturally developed phenomenon seemed to remove it from rational endeavors of invention. Children, as Lemoine pointed out, were very creative in acquiring language, but did not create the language of their environment: "For example, every new-born child does not create the language of its country in its entirety, but because we do not observe them closely enough, we ignore all the creativity this little creature uses during its first education."[48]

The understanding of language as a marker of the human species placed man squarely within the animal kingdom, but also clearly placed him at the top of the assumed hierarchies of that kingdom. It also placed some humans above others in the speech hierarchy: the crying baby—though a precious prefiguration of human ingenuity—did not quite count. It is only after the origin of language within its own existence that it would become a full-fledged member of the articulate/speaking species.

Access to a so-called civilized, articulated language was believed to therefore have a direct, civilizing effect on its learners. The visual language used by sign language users was generally seen as simplistic and only capable of communicating concrete and tangible things, and similar assumptions existed about several non-European languages.[49] In order to be able to express the invisible, such as the basic tenets of the Christian faith, these speakers would need to learn the words used by (Western) educators and missionaries. While the representation of language acquisition as a gateway to full humanity may appear hyperbolic (both sign language users and ethnic others were understood to be part of the same species, after all), the connections drawn between articulate language and emotional abilities and expression show just how fundamental speech was thought to be in the development of an individual.[50] At issue is not the notion that is now ubiquitous in pop psychology that one should learn to express emotions and that talking about one's inner life is somehow therapeutic.[51] Rather, the belief that spoken language could express emotion in a more sophisticated (i.e., more evolved) way than instinctive, animalistic utterances such as cries, grunts, and snarls was read as an indication that possessing articulate language was a sign of a richer, more nuanced inner life. Or, conversely, that those who possessed complex emotions had developed a communicative system to match. Articulate language, as a system of signs to communicate one's inner thoughts to another, was proof of the speaker's place in the grand scheme of evolution.

Expressing Difference

The signs that made up articulate language were linguistic, of course, but also acoustic—and therefore dependent on vocal and auditory qualities or skills, and these needed as much education and polish as the rules and conventions of language. Elocutionist and educator Lewis Monroe even went as far as to compare stress in phonetics to dynamics in music: "The median stress corresponds to the swell in music. It is used in the outpouring of tranquil and fervent emotions; and is specially appropriate in poetic expression. Its effect on the ear is more marked on the emphatic words, but it requires in the whole sentence a certain smoothness. The words are poured, as it were, in a continuous stream. The whole movement is gliding and graceful, not broken and jerky."[52]

The vocal repertoire with which emotions were expressed predated the discovery of the origin of language: in painting, for example, emotional expressions were pictured along largely agreed-upon lines. Historians of emotions have shown how Charles Le Brun's studies of facial expressions can tell us about emotional practice in the early modern period.[53] Likewise, the theatrical stage had developed a range of artful ways to communicate passions and emotions to the audience. These theatrical "rules" were now reinterpreted as naturally evolved (rather than artfully cultivated) expressions of universal emotions. As a result, the voice was no longer an instrument that could be played skillfully to express emotions at will, but was increasingly heard as a revealer of how emotions were genuinely felt. According to Theodore Schmauk, "the variations in pitch and force form an elementary alphabet of expression which can be read at once by any human being."[54] Schmauk then proceeds to give numerous examples of this alphabet of expression in action:

> The soul under the influence of caution, secrecy, fear or surprise, will use the whisper or aspirate quality. If it feels remorse, horror, dread, or deep solemnity, it will express itself by the pectoral quality. Dislike, ill-humor, rage and the ferocity of revenge will come out in a guttural quality. The harsh, discordant quality of an enraged soul, comes from a compressed throat. The steely, bitter tone of hatred is due to the rigidity of the vocal cords; the shriek of terror, to their unnatural tension; the expression of deep solemnity to their relaxation. The genial emotions, serenity, love, tenderness, pity, are expressed in pure quality; but the intense emotions, are aspirate in their vocal expressions.[55]

The particular sounds associated with different emotions were thereby attributed to purely corporeal instinct. Although the ability to manipulate one's voice to call forth affective sounds that do not correspond to any real inner feelings was still recognized, this was increasingly distrusted. This was true in politics, for example, where more dramatic modes of expression would increasingly make way for a subdued style that was heard as more authentic, but also on the stage, where actors would eventually be encouraged to inhabit, rather than put on, a character.[56] Frederick Helmore implored his readers to practice their voices precisely for this reason: "It must be so completely under control as to be capable of expressing every emotion of the soul, whether of joy or sorrow, admiration or disgust, pity or contempt, courage or fear, praise or prayer. Then, and then only does the speaker really speak, then only does the reader read, the pleader plead, or the preacher preach."[57] More expressive still than the voice in speech—but also even more precariously tethered to genuine emotion—was the singing voice. There was a long-standing tradition according to which song (including the motherly lullaby) was a more direct, more instinctive way to communicate emotion than rational speech. The discourse on the affective directness of song and its gendered connotations is impossible to date with any precision. In some respects, the late nineteenth century would become a culmination point for this kind of dualistic thinking about both gender and voice that relegated women's melodious voices to the nursery and men's rational ones to the professional and political sphere. However, just as the political world was not as devoid of emotional voices and female actors as it may seem at first sight,[58] music and singing were not unilaterally linked to femininity, and their role as an emotional conduit was riddled with tension. Like actors, singers onstage were known to cultivate the emotional expressivity of their voices, which created more ambiguity for both singers and listeners to come to terms with. In the same way that breathing, posture, tone-placing, and so on were clearly taught as techniques which were then touted as natural, singers were instructed on how to express emotions that were represented as genuine and intuitively voiced.

Like the vocal repertoires of emotion available to actors, these instructions were hardly new—and, in fact, the belief that emotions and their expressions were natural practices of the species, and therefore universal, precluded any conscious innovation in this regard. After all, if emotions were so widely shared and easily readable among humans, artificially manufacturing new expressions would only serve to alienate audiences. Quite apart from the reformulation and republication of instructions on the expression of emotion by new authors, who couched their writings in modern scientific terms, it is likely that singers also continued to refer to older masters and their advice on the musical expression of emotions. William Shakespeare's *Art of Singing* suggested that the time-honored approaches of the old Italian school rested on a connection drawn between facial and vocal expressions: "The type of expression considered by the old Italian school as the highest test of the freedom of the face, tongue and throat, was that of smiling. The voice must be capable of conveying a smile, and that sound of happiness which is most grateful to those endeared to us must be at our command in every tone we possess, or we cannot sing!"[59]

Likewise, more collective applications of conventions regarding the vocal expression of emotions based on older—largely eighteenth-century—understandings of the nature of different peoples remained in use and in circulation. Transplanted onto newer political configurations, the theories of climatological effects on character largely held their appeal despite the increasing influence of the evolution-based explanatory models described above, which largely pointed toward universally human emotional expression. It was contended that in addition to the environmental impact on the lungs and voice box, different national constellations compelled people to experience and feel different emotions, which were then expressed in an evolutionarily evolved, natural manner. This, too, was connected to understandings of the distribution of different levels of sophistication and civilization. The older theories that took the (political) center of Europe to be "le juste milieu" for the development of a healthy voice owing to its temperate climate were easily understood as evolutionary tales. England, France, and Germany, with their mild or middling climates, could be believed to have offered the ideal breeding ground for vocally and emotionally more developed humans. Discussions on the origin of language did not pose the question of where this origin lay in the world (or when it had taken place). However, the developmental narrative of emotional growth was told as a parallel to the narrative of Western children growing up in nuclear families, and societies growing into industrialized and literate societies, thus suggesting a geopolitical locus for superior evolution.

Overall, these shifts did little to change existing stereotypes. By the end of the nineteenth century, Italians were still generally represented as naturally superior singers; the French were known to have excellent theatrical voices (a notion strengthened by the rising star of Sarah Bernhardt, whose voice was at the heart of her fame); the English were believed to excel at political argumentation (and the growing empire seemed to underline their powers of persuasion); and Germans dominated scientific sociability. Political and biological categorizations of the world and its inhabitants had wrought some change, though, and vocal stereotyping could very easily be mobilized to support new racialized interpretations of difference. Both scientific essays and educational manuals taught their readers to hear national differences in vocal expression and understand those national specificities as evolved, natural, and therefore automatically applicable.

More than to climate, national characteristics in vocal articulation were now connected to language, and although language, too, had been heard as a marker of geographically defined characteristics, the evolutionary understanding of the origin and development of language brought about subtle changes in how national character and language were connected. Older explanations of the connection between national vocal characteristics and a country's associated languages had rested on climatological and environmental understandings of speech. These remained current in the popular imagination throughout the nineteenth century. The idea that vocal practice was a matter of mobilizing air, for example, resurfaced in nonspecialist texts long after it had been thoroughly eradicated in expert circles. The ammoniaphone, a small instrument used to inhale ammonia, for example, was sold on the apparently still plausible message that Italian environments created Italian voices, and that the assumed natural affinity of Italians with bel canto was due to the presence of high amounts of ammonia in the air. Advertised as a "great wonder of the age" with recommendations by both speakers (such as William Gladstone) and singers, the ammoniaphone owed its existence to the enduring influence of eighteenth-century climatological theories on the popular imagination.[60]

The analysis of Italian air as ammonia-holding also mimicked more modern scientific vocabularies in order to sell the product to a late nineteenth-century audience that had learned to assign authority to scientific research but was not necessarily familiar with its details or precise methods. Specialists in the field were largely dismissive of the device: both elocutionists and experts on the singing voice insisted on the cultivation of a physical development of speech and song as a process of civilization rather than a simple strengthening of the body. It is no coincidence, perhaps, that the move toward a neurological or psychological approach to the voice—and the resulting rapprochement between research on the vocal and the linguistic—coincided with the rise and spread of standardized languages. As French, German, and English were defined in grammatical and pronunciation guides, some regional variations were fitted into the straitjacket of a standardized national sound of speech while numerous dialects were put away as merely local, denying their place in the new national order.[61] Next to numerous educational and self-help manuals encouraging the cultivation of a proper voice, similar manuals appeared to encourage the cultivation of a proper pronunciation. Singing teacher Julius Hey's *Art of Speaking* (*Die Kunst der Sprache)*, for example, became a standard reference for the proper pronunciation of German. It was reissued and re-edited numerous times and retains its reputation even today.[62]

By representing dialects as speech impediments, standardized language was identified as not only the correct way, but also the only healthy way to speak. Pathologizing regional and local idioms and pronunciations turned these local modes of speech into "invaders" of the true language, strangers to the national world of sound and therefore potentially injurious to the health of the nation itself.[63] In doing so, proponents of standardized language and speech connected vocal health and propriety to citizenship—a category that was exclusive by its very nature. Most standardized languages were based on the idiom of one particular region, which was thereby turned into the cultural (and sometimes also political) capital of the nation. The Speech Archives (Archives de la Parole) at the Sorbonne in Paris, for example, founded by linguist Ferdinand Brunot in 1911, distinguished between "speech with a fine quality, impeccable rhythm and pure accent" and "speech marked by suburban or provincial accents"—in other words, nonstandard speech was heard as accented and impure, somehow bending proper French out of shape.[64] As notions spread that a language like Nissart could no longer be seen as French, Sächsisch did not count as German, and Geordie was not proper English, not all inhabitants of those regions necessarily became seen as second-class or failing citizens. Rather, those who made up the lower strata of society almost anywhere in the nation became insufficiently national in their speech. The upper classes had always been mobile—they were not bound to any local style of speech per se, as they conversed fluently in the different regions where they resided or traveled, doing business across long distances or even traveling internationally. What became suspicious once dialects were seen as non-national was the language of farmers or factory workers: those who could not afford the kind of mobility that would have familiarized them with the language of the middle and upper classes of the capital, and who had grown up exclusively with very local sounds and slangs.

By the time the process of standardizing language (which had started around the seventeenth or eighteenth century) reached its peak at the end of the nineteenth century, these local modes of speech could also be slotted into the logic of linguistic evolution, within which they appeared as uncivilized:[65] not quite as low-ranked as non-European languages (especially those including signs or clicks), but seen as not fully developed nevertheless. Much like "others" from elsewhere, dialect speakers were categorized as somewhat wild, their speech as an incomplete device, only fit to communicate the less sophisticated emotional world and lower intelligence of its speakers. At the same time, emerging theories of degeneration, especially in industrialized cities, showed anxieties over the particular direction evolution had taken in environments devoid of proper hygiene, education, or (moral) rules. Sounds that were heard as uncivilized, such as the Zulu language, could still be heard as a sign of unspoiled wilderness, a step on the way toward European perfection (and therefore a piece of the evolutionary puzzle, proving that language and the people who spoke it were the result of a process of selection). The perceived equally uncivilized braying of the lower classes cast doubt on the linearity of that process, and on the general belief that education could help cultivate natural, human characteristics in those who appeared as less than human. In more flattering terms, such beliefs about uncivilized voices were also supported by narratives of naturally brilliant singers, the lower-class diamonds whose sparkling voices seemed to emerge naturally from unformed throats. Enrico Caruso, for example, stressed his urban working-class background to underline his own exceptional status as one who needed little training to become a unique and celebrated tenor.[66] The image of the unconstrained and therefore healthy voice of the lower classes was also upheld by specialists on stuttering, who insisted that the impediment was connected to stress that could only be experienced by the (successful) modern man, and was therefore absent in others.

Although the seats of both language and intelligence were believed to be located in the brain—an organ that drew increasingly more medical and cultural attention—the psychological analysis of the origin of language also supported the ongoing belief that numerous personal characteristics were refracted through vocal utterances and could thus be heard. More importantly, what medical and educational professionals taught themselves to hear were not just the personal characteristics of each individual in their care, but rather vocal markers of collective identity. These new approaches therefore also opened up new ways to read vocal utterances as expressions of belonging and community, and conversely, to think of vocalization as a way to support the formation and strength of communities. The education of the voices of the people was therefore a practice of great importance, as it would decide on the health of the nation. The upper classes sent their sons to universities, where they would learn to speak properly in debate clubs,[67] and their daughters to private tutors for singing classes.[68] The middle classes were the main audience for the new phenomena of therapy and the self-help manual, but the majority of the "people" depended on mass education for their vocal instruction. In primary schools throughout Europe, instruction in music—and song in particular—would become an important pillar of the education of the people. Simple notation systems were devised to make sheet music more accessible to children or the less educated, numerous songbooks were compiled to be used in the classroom, and teachers of young children would increasingly be taught to play musical instruments to accompany and guide their charges in song. Although serious professional and cultural work was put into the development of a musical curriculum in mass schooling, this was generally not presented as a particularly difficult endeavor: singing had always been associated with children and the nursery, changes to the practices of singing appeared gradually rather than through unilateral policy changes, and, above all, singing was seen as a very cheap strategy to be implemented in support of children's health. As Leo Kofler pointed out, "the voice is the most popular musical instrument . . . because it is by far the cheapest musical instrument; it cannot be bought, it is always a gift."[69] The same tactics, albeit to a lesser degree depending on the nation in question, could be applied to adults of the working class. Choirs, like the German Gesangvereine, the French Orphéons, and later numerous groups of laborers who formed choirs at the factories or mines, sprang up alongside a movement to spread the skill of sight-reading among those who would never learn music in a drawing room.[70]

As with the singing in schools, these societies had ambitious goals. Health was presented as one of them, morality and the civilization of its members as another. Choral societies often set great store by discipline, for example, and had specific rules and sanctions about the use of coarse language, "order and reverence," and the avoidance of vulgarity.[71] The sociability within those choirs was therefore a means to influence their members' habits and convictions as well. This was effectuated through the mere practice of song as well as the nature/content of the particular music they chose to perform. By its very nature, choral singing lent itself well to the representation (and experience) of collectivity, as had already been clear in the earlier part of the nineteenth century in writings such as those of Joseph Mainzer and Laurent de Rillé, who pointed out that choral singing could strengthen the cohesion of the local, regional, or national "tribe." In the second half of the nineteenth century, this tribe was to be imagined on a much grander and more fixed scale. Although regions continued to play their part in the organization of choral societies, the national level of belonging and sociability was the one to be practiced, learned, and celebrated in song.[72] Different nations implemented this strategy in different ways, reflecting differences in existing choral traditions and educational systems. In England, where highly trained male choristers remained the norm for more formal choirs (and the inclusion of female altos was considered "objectionable"),[73] sol-fa-ing became a democratizing musical movement for the people.[74] In Germany, the Gesangvereine had a distinctly upper- and middle-class character, but the (male) choir would also become a "popular" means of sociability, and in France, the somewhat bureaucratic system of the Orphéons sprung up from a country made up of regions and *arrondissements*.

To the musicians and educators who were active in these choral movements, national differences appeared to indicate an implicit competition—or hierarchy—between nations in matters of musical education and song. As in the fields of laryngology or linguistic neurology, the network of experts on choral singing and musical education was an international one, with authors generally reading in different languages and showing an interest in various schools and systems across Europe (but mainly its Western center: Germany, France, England, and Italy). However, information was exchanged less freely on this subject. Laryngologists and neurologists may have skirmished over priority of invention, but these moments of competition were between schools of thought or individual heroes: neither the invention of the laryngoscope nor the discovery of the speech center were claimed as national victories, and the inventors on whom consensus eventually fell were thoroughly international figures, traveling around Europe to be celebrated for their knowledge and to exchange thoughts with colleagues. Because mass education in music and song was designed to further the health of the nation, the international outlook of those developing such national programs was less cosmopolitan in its approach. When they were learning about successful attempts elsewhere, this knowledge was adopted and appropriated to one's own national context, much as scientists had adopted the use of the laryngoscope for other fields. But foreign achievements were also used to reflect on how one's own nation measured up, internationally, and to lament the lack of enthusiasm for choral singing or music education at home. (Curiously, virtually all authors looked at their own nation in despair, invariably pointing out to their readers how horrid it was to have to admit to inferiority in this matter in comparison to their neighbors.)

Despite this insistence on national specificity and hierarchy, the discourse on choral singing and musical education as well as their main goals was remarkably similar across different countries. Aside from physical and moral health, singing was seen as a way to prepare children's bodies and minds for adulthood. Much like the gymnastics classes that were rising in importance around the same time,[75] music lessons in schools were (systematically) applying principles of modern medical and psychological knowledge. And also like the gymnastics lessons, the nature, goals, and benefits of singing were imagined to be thoroughly gendered. Most importantly, singing would allow girls to practice the movements, sounds, and emotional expression that would later constitute their practice of motherhood. This preparation for adult life was not (only) designed to be to the benefit of girls and women themselves, but mainly to more effectively cast them in the role of bearers and rearers of the nation. The future of the country and its people was thus laid in the hands of young women, who therefore had to be educated to properly fulfil their duty to the "tribe." Learning to sing played a part in this not only because it would keep women healthy, but also because it made them more effective messengers for the fatherland. As Julien Tiersot noted, for example, nobody doubted "that choral song is an incomparable agent of moral civilization and national education, when the language of music expresses, in its own words, the generous sentiments to which every man's heart responds."[76]

And for such choirs of young children to exist, healthy vocalization needed to be encouraged from birth, and defects in voice production were to be—as Morrell Mackenzie put it rather grimly—"strangled in the cradle."[77] The image that prevailed in discourses on musical education was, ultimately, not the healthy young patriot ready to defend the fatherland while robustly intoning a rousing song, but rather a young woman rocking a child with a lullaby on her lips. The romantic esthetic of that image—or just its sentimental appeal—was central to its power and helped spread a much more pragmatic message: that the sound of a healthy voice indicated a healthy citizenry.[78] National health, according to such an image, was passed on from generation to generation. The notion of melodious motherhood cast the practice of singing in the role of an evolutionary tool—as one characteristic that had marked out fitness and was therefore to be passed along. And although song was (and had to be) cultivated to truly contribute to the well-being of the nation, its strong connection to motherhood also put song at the heart of reproduction. The age-old connection between the feminine and song was repurposed and transliterated, at the same time as material, anatomical explanations of speech and the voice made way for the neurological and psychological understandings of speech. Neurology and psychology would not only place the origin of language in the cradle of each child, but would also imagine that process as a parallel to the origin of speech itself.

Moreover, the practice of passing on vocal health from one generation to the next was believed to contribute to the creation of national (or at least geopolitically defined) voices. Authors across genres and specializations were convinced that national or regional identity could be heard not only in language or dialect, but in the voice itself. Italians, in particular, were believed to produce a distinctive, recognizable sound. According to Georg Hermann von Meyer, the beauty that was attributed to Italian speech was entirely due to qualities of voice and tone: "Thus we often hear the remark that Italian is a beautiful language, because it has such a musical sound, when evidently only the voice-quality of the Italian language is implied."[79]

And while most authors conceded the superiority of the Italian voice, connections between national sounds and vocal health also led to ambitions and efforts to "make" the national voice according to patriotically informed ideals of vocal beauty and strength, as expressed, for example, by the aforementioned Joseph Mainzer, who presented vocal education almost as a matter of national competition.

Doing so forged a connection between individual maturation and the growth of the nation—a connection that would become increasingly important over the course of the nineteenth century. Recent research on the histories of experience and emotion have pointed toward the important role individual practices and feelings played in the realization of politically inspired communities like the nation.[80] It comes as no surprise that authors writing for a middle-class audience expressed similar attachments to a brand of propriety and good behavior that was valued as a sign of citizenship, belonging, and love of the nation. But connecting the acoustic qualities of the voice to such patriotic affective practices and experiences does teach us something about the practical manner in which such ideals were given shape and articulated. It also gives us a sense of how individual practices and collective ideology were brought together—and remained tethered. The highly personal but intrinsically intersubjective practices of speaking and singing helped forge such communities at a very practical level.

This was conceptualized as a long-term process even by contemporary observers: much though belonging to a nation was seen as a birthright, the ability to articulate it properly was very much seen as the result of sustained practice and experience. Older notions of climatological influence on the body, the organs, and its performance guided these narratives in often counterintuitive ways that became increasingly more difficult to square with emerging ideas on biology, evolution, and the development of individual bodies as well as whole populations. According to an early nineteenth-century essay on theatrical performance, for example,

> It has been noted that voices and pronunciation are more guttural in hot countries, where articulation depends more precisely on the action of the posterior parts of the organ, whereas in cold countries all the articulations are labial and are due evidently to the action of the anterior parts of the vocal instrument. It is easy to see the reason for this effect. A mild temperature invites naturally to dilate the organ of the voice and to

> allow it to deploy the full extent of the movement of which it is capable, whereas the influence of cold keeps it in a state of continual constraint and contraction; so that among the inhabitants of cold countries, it is on the most external parts of the organ that all the effort and all the action of the voice must bear; which supports the theory I advance.[81]

This attention to small differences between European nations also prefigured what would, eventually, become a distinctly imperial approach to proper speech that was reserved for speakers of a small number of languages deemed sufficiently civilized, and for voice boxes formed in temperate climates. Such limitations were based on older, supposed connections between regional character or morals, the climate, bodies, and language. In an essay that seems like an earlier iteration of Alexander Melville Bell's project to represent "all human sounds" on paper, French prosodist Faure Lacloze described what he understood as a causal connection between climate, character, and sound:

> harsh sounds and multiple consonants in the same syllable seem to announce stubborn, thinking, serious and phlegmatic nations, like the Germans; civilized, heavy and silent nations like the English and the Dutch, who live in a cold climate, will probably sound strained and lazy; gay, sweet and voluptuous nations with a pleasant climate, like the Italians or the Tahitians, will have many sonorous, smiling voices; grave and angry nations, like the savages of America, Spaniards, Turks, etc., whatever the climate, will have full and serious sounds; while civilized, humane, industrious peoples, and above all conversationalists like the French, in a temperate climate, have a large number of sounds, but do not use harsh sounds nor do they frequently use low pitches or double and triple consonants.[82]

Despite their interest in comparative biology and big questions of humanity and its special status, very few authors on the human voice showed much of an interest in human voices outside of the small world of middle-class Western Europe. When non-Europeans make an appearance in their work, they often do so as essentially speechless figures in need of instruction in order to learn to make themselves understood. This was for example one of the envisioned functions of Bell's "visible alphabet," which would not only allow people he identified as deaf-mutes to acquire speech, but would also facilitate "The speedy diffusion of the language of a mother country throughout the most widely separated colonies."[83] This notion of the speechlessness of the non-European often took exceedingly literal forms. For example, John Wyllie recounts a narrative wherein the utterances of what appears to have been a Sàmi woman were interpreted as not only not linguistic in what was considered the normal European way, but not as voiced at all:

> A deaf and dumb lad, named Collins, is taken to see some Laplanders, who were carried about to be exhibited, and writes thus to his fellow-pupils about the Lapland woman: "Mr. Joseph Humphreys told me to speak to her by signs, and she understood me. When Cunningham was with me, asking Lapland woman, and she frowned at him and me. She did not know we were deaf and dumb, but afterwards she knew that we were deaf and dumb. Then she spoke to us about reindeers and elks, and smiled at us much."[84]

The story reveals just how closely notions of health, civilization, and evolution came to be related to each other by the end of the nineteenth century:[85] the assumed mutual understanding between a "deaf and dumb lad" and an ethnically othered adult was predicated on the perception of a shared lack of sophistication. "The gesture language of deaf-mutes is so closely allied to that employed by uncivilized races," Wyllie contended, "that deaf-mutes and savages are easily able to converse with each other by means of gestures."[86]

Much like the voices of the (dialect-speaking, rough around the edges) lower classes, those of ethnic others were categorized as far removed from propriety and civilization—and were thus denied audibility. Recent research on, for example, Black vocality in the United States and Western Europe has demonstrated the enduring effects of these practices of categorization: on the one hand, Black voices onstage were starkly on display in contexts of segregation (which makes it possible to study documents of their existence today), while on the other hand, outside of these particular contexts of artistry, the "sonic color line" made the day-to-day speech of Black individuals all but invisible, resulting in an enduring lack of documentation.[87] While the assertion that the "Lapland woman" did not speak at all is a particularly extreme example of this refusal to hear the voices of ethnic others, it is symbolic of a more widespread practice of failing to hear speech in the vocal utterances of large numbers of (non-European) people. The well-documented failure/refusal to understand the utterances of Indigenous and colonized peoples as meaningful speech had far-ranging political and legal consequences—as for example Dylan Robinson and David Le Breton have shown through their analyses of listening and silence in the context of the courtroom.[88] What the pathologizing vocabulary around non-Western speech shows is just how deeply material the results of this lack of audibility could be—resulting not only in a loss of sovereignty and individual autonomy within the legal and political frameworks of the modern (imperial) state, but in a loss of physical integrity altogether. As Joanna Bourke has pointed out in *What It Means to Be Human*, there was a violent quality to the way in which speech was used to distinguish humanity from its others.[89]

The pathologization of "other" speech went hand in hand with its infantilization, and both processes were indicative of the profound lack of autonomy imposed upon those whose speech was not heard.[90] Bell's ambitions for visible speech, for all their apparent emancipatory goals, underpinned this discourse in which disability, ethnic difference, and childishness were intrinsically connected. Likewise, all three were supposed to be overcome through the careful application of pedagogical and scientific methods. The line between education and cure was thin, as infantilization and pathologization were applied to similar afflictions or defects and largely confused with each other—lisping, psellism, and cluttering could all be heard as either disease or temporary childish incompetence.

The language of infantilization and the drive to teach ethnic others to speak properly would, eventually, also aid in the reconceptualization of norms of proper speech in the context of emerging ideas around evolution. As the processes of growing into humanness and adulthood were thought to resemble each other, the various methods and strategies that had been developed in educational contexts for (mostly) middle-class children could also be applied to cure the inferior speech of racialized and pathologized others. The rise of a more therapeutic approach toward healthy vocalization was, in other words, not an isolated issue of medical innovation or professional change. It occurred in a wider social and cultural context in which vocal articulations of self served to co-construct the daily practice of empire, and, conversely, in which understandings of empire co-defined the edges of humanity itself. If, as Hedwig Richter puts it, freedom can be rendered as being protected from physical punishment,[91] belonging to the increasingly biologically defined category of human would render a person as someone whose speech is heard as curable.

Or, to put it differently, while the pathologization and infantilization of the speech of ethnic others put them in a position of evolutionary inferiority, it gradually removed them from a position of servitude. On the sliding scale of evolution, the firm distinction between man and beast became fluid. This opened the doors to crass racial hierarchies and eugenic thoughts, but also allowed for a redefinition of humanity (as among the animals) as well, and, counterintuitively, made some ethnic others' voices more audible than they had been before. The assumed bestial nature of ethnic others, in an evolutionary world, marked them as distant kin rather than as complete strangers. And the same would apply to numerous animals—especially those capable of speech. As Bruce Boehrer has pointed out in his cultural history of the parrot, the speaking birds had been likened to servants in earlier periods,[92] but in the nineteenth century the parrot became a focal point for studies of comparative biology, positing the acquisition of speech on an evolutionary scale in which human adults represented the peak and various others were positioned on lower levels. Throughout the nineteenth century, parrots were compared to children—and vice versa. Much like so-called "idiots," children appear in these stories as lesser humans, whose parrot-like behavior is the subject of interest, but also of scorn. The speech of both birds and children was understood as "mere reflex."[93] Infants' transformation of cries and murmurs into syllables, such a miracle to young parents, was taken with an almost palpable shrug by scientists as something that "even takes place in parrots."[94] Advice for the education of children, finally, included notes to avoid repeating sounds "à tort et à travers" like a parrot. Rather than marveling at the capacity of a bird to produce human-like sounds, these educators despaired of the animalistic nature of young children. According to a reading exercise in a French manual of 1880 by Arthur Chervin, "Nothing is quite as tiring, in company, as blathering children who constantly interrupt serious conversation. They are, quite rightly, compared to parrots whose incessant chattering is amusing for a moment, but quickly becomes insufferable."[95]

An Abundance of Parrots

Comparative biologists' increasing reliance on Darwinian notions of interspecies kinship brought the spaces of overlap in human and animal voices into sharp relief. As Thomas Dixon has noted, the image of the Great Chain of Being placed man at the summit of a natural hierarchy that was further strengthened by the human capacity to cultivate his own talents. This superiority entailed "the intellectual ability to reflect on one's past behavior, the linguistic ability to communicate and understand communal approval and disapproval, and the power of hereditary transmission of habits and instincts."[96] But physiologists in their laboratories—and often also with impressive domestic menageries—kept coming across animal behavior that questioned the uniqueness of the two latter abilities. The talent for mimicry and suitability for training of some animals had of course long been known and exploited for industry and entertainment. But at the end of the nineteenth century, numerous stories of animals with intelligence surfaced in scientific work, often couched in the language of neurology or psychology. In 1892, for example, retired physician Abraham Netter published his *La parole intérieur et l'âme* and included "the story of the rabbit, told by professor Laborde."[97] Netter structured the account as an almost fairy-tale narrative, mimicking the oral heritage of storytelling: "Do you know the story of M. Laborde's rabbit? It is entirely true and authentic: 'The rabbit in question was legendary in the old laboratory of physiology: nobody who frequented that establishment between 1877 and 1880 could have forgotten the illustrious critter, who answered to the name "Bertrand."'"[98] Having introduced Bertrand, the story goes on to detail his life in the laboratory, including the experiments he underwent and his love of breadcrumbs, but also his relationships with other animals in the laboratory, his aggression toward strange dogs, and his affection for Professor Laborde, who takes up the narrative as follows:

> He was present as if he wanted to participate, and certainly gave the impression of being interested in my experimental work. His expression and attitude reflected an attitude of curiosity; he would stay on the experiment table for hours, where the object of study, a dog or a rabbit, was held, as if he was observing them with us. But what always seemed of most interest or intrigue to him, were the microscopic studies. I would barely be seated at the microscope, before Bertrand would endeavor to get on the table or unto my shoulder, and he would take the exact same position as the one he had seen me in, pressing his eye unto the microscope's lens.[99]

The personal relationship between these scientists and their laboratory—or pet—animals lends the stories a particularly human touch, but also reveals some of the genuine convictions these men felt for evolutionary theories of kinship. The tendency to take human experience seriously in research (as noted in Chapter 3) extended to these experiments on human-animal affinity, and pet-owning scientists' attention seemed to be drawn quite naturally to the possibility of communicating with animals as a way to explore their neurological and vocal (or even linguistic) abilities. Georges Romanes's menagerie of dogs, monkeys, and birds provides an interesting example. In his exploration of the physiologists' identity construction as a gentleman scientist, Rob Boddice describes Romanes as a self-professed animal lover who valued his relationship with his dogs and did not think twice over sharing his house—to the horror of his wife—with one of his lab animals, a monkey.[100] In fact, experimenting on and sharing experiences with his animals were, as Boddice shows, intimately connected for Romanes, who allowed his pet dog Major to be vivisected precisely because he thought of the animal as kin, and believed that dogs possessed some sort of intelligence.[101]

For an exploration of human-animal communication through vocal practice, I'd like to focus on another member of Romanes's somewhat exotic household: his parrot. Nottidge Macnamara included the bird in a rather long succession of talkative pets who were "acquainted" with various professors.[102] As Macnamara reported,

> when [Romanes's parrot] saw the coachman come for orders, she would at once exclaim "half-past two," the bird having repeatedly heard this order given, had imitated the sound, and associated the words with the man who received the order. This bird at dinner-time had been accustomed to have savoury morsels of food given to her, and had been at these times taught to say, "Give me a bit," which the bird constantly repeated, but only and appropriately at dinner-time. The bird associated the expression with something to eat.[103]

True to his evolutionary perspective, Macnamara interpreted the bird's abilities as a simple form of mental activity, comparable to that of humans but less sophisticated. Like many scientists of his time, he explained the parrot's speech as childlike. This was in keeping with a long tradition (the ancient Greeks commented on parrots' human-like voices as well, and generally represented them as servants or students).[104] By the end of the nineteenth century, however, it became connected to an evolutionary model of neurology. According to Macnamara, parrots not only sounded a bit like a human, they also acquired speech in a way that he expected would remind his readers of human children: "We notice the same faculty [associating expressions with objects] in an infant long before he has attained the power to think or to reason."[105] Even the birds' posture and movements were described in an implicitly anthropomorphic framework, suggesting that parrots paid special attention to human facial expressions while learning: "If we watch a parrot while he is being taught to speak, we notice that the bird turns his head first to one and then the other side, as if striving to catch the sounds which are being repeated to him. The bird is all attention, listening not only to hear the sounds which are being spoken, but also attentively watching every movement of the face of his instructor."[106] It seems that Macnamara had watched a great number of parrots in the process of using or acquiring speech—his lengthy descriptions of some of these birds differ uncomfortably little from the description of Laura Bridgeman, a deaf, mute, and blind girl he later focuses on in *Human Speech*, giving the distinct impression that he thought of talkative birds and mute people as occupying a space at a similar distance from actual human speech.[107]

So what made the words uttered by birds, and studied by biologists, so valuable to laryngologists and speech therapists? And what do their questions and conclusions about parrots tell us about their notions of humanity and voice? Looking at the manuals and research produced in the nineteenth century, we come, like Robinson Crusoe, across an "abundance of parrots" that were not valued because of their exotic nature (as was the case for most parrots before the nineteenth century) or even their connection to the politics of empire and global trade, but exclusively for their speech.[108] Quite tellingly, none of the parrots mentioned in these books are described in terms of plumage, size, or color. They were interesting to these scientists because of the evolutionary puzzle they presented. Here are ostensibly lower animals, and yet they seem to have at least some affinity with what had always been called *the* distinguishing characteristic of humans: speech. Frederik Mott, in his early twentieth-century monograph *The Brain and the Voice in Speech and Song,* relied on "progressive evolution theory" to support this hierarchical view. "Articulate language," he notes, "is the result of an elaboration in the long procession of ages in which there occurred three stages—the cry, vocalisation, and articulation."[109] Parrots necessarily take up an uncomfortable place in that chronology, because even though "it is intelligence that has converted the vocal instrument into the speaking instrument" in humans,[110] these birds seem to have access to very similar instruments. To Mott, their speech seemed even superior to the language of some people, whose gestures, clicks, and "inarticulate" sounds were interpreted as "primordial" instances of human speech.[111]

The difference between animal and human speech, according to Mott, was the use of "judgement."[112] His interpretation thereby dovetailed with that of Macnamara, who typified parrots' speech as an "automatic reflex process" in which there are "indications of consciousness," but not of the intelligence that can be ascribed to humans and higher animals.[113] What made the parrot so interesting, then, was not so much that it seemed to be able to speak, but that the particular lacunae in its capacity for speech highlighted human uniqueness and helped in understanding human disability. The comparison between animals and humans was at its most revealing when the difference between them seemed blurry, when

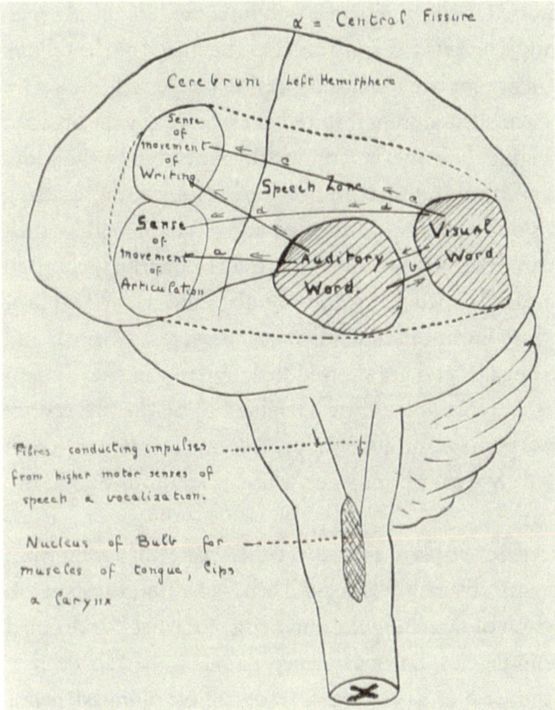

Figure 8. Frederick Mott, *The Brain and the Voice in Speech and Song*, London: Harper and Brothers, 1910. Courtesy of King's College London.

the human-like animal seemed superior to the not-quite-fully-human person. Albert Lemoine posed the question of comparison perhaps the most clearly in his *Physionomie de la parole* in 1865: "What is missing, then, in the man who is condemned to eternal mutism unless some miracle unfolds inside or outside him? Does he lack a voice, like fish, the ability to articulate sounds, like the ape, reason, like the parrot, or does he lack hearing like the deaf?"[114] And thus there was hierarchy within pathology as there was within humanity—a hierarchy that could be expressed by referring to a metaphoric bestiary of impediments. The parrot, in this system, aided in recognizing and defining intelligence, or its absence: "In the end it is the idiot whose imbecilic reason makes him resemble the babbling parrot, not the deaf man."[115] For scientists around the turn of the twentieth century, the parrot's particular form of consciousness was of interest because of the neurological processes it exhibited: the practice of teaching a parrot to speak told biologists something about "lower" brains.

This turn toward the brain makes scientific comparisons between parrot and human voices appear as a niche scientific pursuit: a shift from natural philosophies of voice interested in the uniqueness of humans to biological explanations of the vocal evolution of humans and animals. The image of the talking parrot, however, and its salience for late nineteenth-century experts, is indicative of a much more complex sociocultural shift, inspired not so much by the brilliant insights of individuals changing their own emerging disciplines, but by the wider practices and experiences of both industrialization and colonization. Existing social hierarchies and divisions were both disrupted and strengthened by such changes—putting the urban poor in a position of extreme infantility, children and youth in a place of increased dependency, and colonial subjects in the role of the other against which white European humanity was measured. Manuals and scientific treatises on vocal health and art showed little interest in these infantilized and silenced others: because speech was imagined to be the preserve of so-called civilized speakers, perhaps, or even simply because so many of them explicitly addressed a relatively well-off audience who could afford books and the leisure time to read them.

Nevertheless, ethnicized, lower-class, and other others were not absent from their work. Narratives of exotic, brutish, rough, and otherwise improper speech served the important function of sketching the outer edges of what was deemed acceptable for the white, middle-class European voice. As mentioned above, perceived silence fell outside those edges, as did a number of audible sounds depending on volume or pitch. Falsetto sounds, particularly in men, could, for example, be described as inherently wrong or false. According to William Gardiner, "Its high and piercing cry is rather the language of imposture than sincerity. In the voices of men the voce de testa is sometimes termed a falsetto, or feigned voice, the tone of which is similar to the constrained effect of over-bowing an organ pipe, or a flute. This fictitious voice is now abandoned by composers of the present day, as being devoid of strength and expression."[116] Especially when it came to ethnicized others, however, very few of these narratives were based on actual observation. They also differed significantly from descriptions of middle-class pathologies or healthy practices in their lack of precision: the mirror image of the carefully drawn picture of the stammerer whose profession, background, and life story were recounted to contextualize his condition was not in fact an equally complex individual in another, less industrialized, world; it was the collectivized savage on the plains of Africa, or a conglomerate of Laplanders signing to represent the overall silence of the North.

Whereas the language of exploration, and the intellectual and cultural tools wrought by colonization, were in full use in these handbooks, the influence of Europeans' imagination of themselves as explorers and settlers was neither recognized nor discussed—and it remains difficult to do so now. This chapter has much more to say about small differences among middle-class Europeans not only because they were the subject and audience of scientific and pedagogical texts, but also because authors were actively invested in magnifying these internal differences while downplaying the complexity of vocal behavior and health in the rest of the world. Equating good speech with humanity and its unique and superior status among other creatures implied an investment in certain types of voices and certain sounds of speech over others. The benefits of increasingly systematic and precise research on vocal anatomy, linguistic practice, and the development of both were therefore reserved for bodies deemed fit to project such precious voices.

INTERMEZZO 4

Simian Sounds

Acoustic technology would, eventually, herald enormous changes in the ways humans understand and interact with sounds, including those of their own and others' voices. Work in sound studies has gone a long way in elucidating those changes, showing how new technologies encouraged the development of new sonic skills,[1] how the new portability of sound changed its relationship to space,[2] and how various methods to reproduce the voice slowly detached voices from living, breathing bodies.[3] As research on technology like the stethoscope and various recording devices has shown,[4] new acoustic technology also drove innovation in a number of scientific disciplines like medicine, psychology, mechanics, and philology. Joeri Bruyninckx, for example, has studied how recordings contributed to the development of new practices of observation in ornithology.[5] Although it would take a while for acoustic technologies to become sufficiently portable for this kind of work, the use of recordings in the study of animal sounds arose quite early—by the end of the nineteenth century. The gramophone, which was mobilized to record so many varieties of human song and speech, would also be used to reimagine speech itself, and to extend its definition beyond the confines of the human throat.

In 1892, Richard Lynch Garner reported on his experimental use of the gramophone in the zoos of New York, Washington, and Cincinnati. In *The Speech of Monkeys*, he recounts how he

> procured a phonograph upon which to record the sounds of the monkeys. I separated two monkeys which had occupied the same cage together for some time . . . I then arranged the phonograph near the cage of the female and by various means induced her to utter a few

> sounds, which were recorded on the cylinder of the phonograph. The machine was then placed near the cage containing the male, and the record repeated to him and his conduct closely studied. He gave evident sighs of recognizing the sounds, and at once began a search for the mysterious monkey doing the talking.[6]

His work was based not only on the assumption that monkeys could communicate with each other but that with enough study and practice, humans could communicate with them too. As he stated in the introduction to his book on monkeys:

> From childhood, I have believed that all kinds of animals have some mode of speech by which they could talk among their own kind, and have often wondered why man had never tried to learn it. I often wondered how it occurred to man to whistle to a horse or dog instead of using some sound more like their own; and even yet I am at a loss to know such a sound has ever become a fixed means of calling these animals. I was not alone in my belief that all animals had some way to make known to others some certain things; but to my mind the means had never been well defined.[7]

His research project on the speech of monkeys elicited "a smile of incredulity" from the "great fathers of science,"[8] but was presented as a serious and systematic endeavor. It was, as Garner phrased it, an experiment wherein "for the first time in the history of language, was the Simian speech reduced to record."[9] Garner describes a number of Simian individuals in his book; most of them are not only mentioned as representatives of their species but also indicated by name: "a brown Cebus whose name is Jokes,"[10] "a little capuchin by the name of Pedro,"[11] etc. At the heart of this study is an understanding of evolution that places monkeys and humans in the same family tree, if cautiously so:

> I do not pretend to know whether man was evolved from ape, or ape from man; whether they are congenetic products of a common authorship, or the masterpieces of two rival authors; but I cannot see in what respect man's identity would be affected, whatever may be the case. . . . That evolution is the mode by which the world was peopled, there is little doubt, but there are many details yet unsettled as to the manner in which this was effected.[12]

And although Garner points to a number of physical, phonetic, and grammatical differences between human and simian speech, he insists on the latter's profound effects on human emotions, and on its intrinsically social nature, noting that "monkeys do not talk when alone, or when it is not necessary to their comfort or pleasure; and while I am aware that their speech is far inferior to human speech, yet in it there is an eloquence that soothes, and a meaning that appeals to the human heart."[13]

In fact, Garner's study of the speech of monkeys reads much like an anthropological study, focusing not only on the morphological features of language, but also on what could be described as cultural practices of different individuals according to ethnicity and gender. One female capuchin, for example, was designated "the Juliet of the Simian tribe" and described in great detail—inscribing her behavior with expectations and interpretations that seem more at home in the Victorian drawing room than in the zoo:

> Dodo would stand erect on her feet, cross her hands on her heart, and in the most touching but graceful manner go through with the most indescribable contortions; she would sway her body from side to side, turn her head in the most coquettish manner, and move her folded hands dramatically, while her face would be adorned with a Simian grin of the first order, and the soft, rich notes of her voice were perfectly musical. She would bend her body into every graceful curve that can be imagined, move her feet with the grace of the minuet, and continue her fervent speech as long as the object of her admiration appeared to be touched by her appeals. Her voice would range from pitch to pitch and from key to key, and, with her arms folded, she would glide across the floor of her cage with the grace of a ballet girl; and I have seen her stand with her eyes fixed upon her keeper, and hold her face in such a position as not to lose sight of him for a moment, and at the same time turn her body entirely around, in her tracks, with the skill which no contortionist has ever attained. During these orations I have observed the little tears standing in the corner of her eyes, which indicated that she herself must have felt what her speech was intended to convey.[14]

Much like early ethnographers, Garner also seems to engage in modes of intercultural communication—without too much regard for his own inability to speak his study-objects' language. Pedro, a small capuchin in Washington, was presented as a particularly active interlocutor and a fount of information about the zoo in question despite his unintelligible language:

> When I would go to record his sounds on the phonograph, I held him in one hand, while he would take the tube in his tiny black hands, hold it close up to his mouth, and talk into it just like a good little boy who knew what to do and how to do it. . . . He was quite fond of the head-keeper and also of the director, but he entertained a great dislike for one of the assistant-keepers, and he has very often told me some very bad things about that man, but I could not understand them. I shall long remember how this dear little monk would cuddle up under my chin, and try so hard to make me understand some sad story which seemed to be the burden of his life.[15]

The book contains a number of ethnic comparisons, too, constructing a hierarchy among monkeys based on largely anthropomorphic criteria, and recognizable through features central to anthropometry (such as cranial style) as well.

Practices of learning and imitation occupy an ambiguous place in this narrative. The case of "Uncle Remus" is a rare one in Garner's work: the Cebus, held in Central Park, had learned to adopt speech-like behavior by imitating a capuchin in an adjoining cage because he wanted to interact with humans. Garner's primary interest went to the kind of behavior he saw as their own, autonomous and authentic, language. Nevertheless, his work also shows signs of what would later become more common in studying the communicative talents of monkeys: their treatment as child-like learners of a behavior that was attributed to a supposedly superior species but that was nevertheless natural to them too. He observed the sign of a shake of the head for "no" in monkeys' acquisition of language, for example, and, having "inquired among mothers who have raised families," declared it a practice shared across races and species. Remus, Pedro, Dodo, and the other "little creatures who have aided me in the study of their native tongue"[16] prefigure the simian faces of twentieth-century research subjects like Nim Chimpsky and Koko the Gorilla.[17] And much like the speaking and signing monkeys of the future, they were seen as holding up a mirror to human behavior, questioning human claims to a unique empathy and kindness. Garner's monkeys subtly showed what another speaking monkey—Kafka's *Rote Peter*—would express much more sharply in the early twentieth century: that the implicit hierarchy of teaching and learning languages across species was as indicative of human vices as it was of potential virtue.[18] As Kafka's monkey narrator explains, "It was so easy to imitate these people. I could already spit on the first day. We used to spit in each other's faces. The only difference was that I licked my face clean afterwards. They did not."[19] Or as Garner put it more kindly, "A knowledge of their language cannot injure man, and may conduce to the good of others, because it would lessen man's selfishness, widen his mercy, and restrain his cruelty."[20]

At the same time, these Simian speakers also referred back to older traditions of human-animal cohabitation and collaboration—or of species companionship, as we may now call it. As Garner pointed out, dogs have had a long history of coexistence with humans. And they are hardly the only animals to share large swaths of human history. Horses, taking part in warfare, farmwork, and fashionable social life in a number of ways, have even been shown to be so closely intertwined with humans' lives that we may need to think of at least some histories as centauric more than strictly human.[21] The notion that animals possess communicative qualities long predates the nineteenth century, and questions concerning their ability to feel or their possession of souls resurface throughout history.

What is particular to modern considerations of animal-human coexistence and companionship is the reliance on tangible and audible vibrations captured in the new tools and technologies of comparative biology. Victorian scientists were less concerned with animals' souls and more with their potential intelligence. Speech provided what seemed like a clear avenue to gain knowledge about that intelligence because it gave the abstract quality a material, embodied articulation. Lodged in both the organs and the brain, practices of speech could delineate the differences between human and primate modes of communication . . . but could also point toward similarities or even overlap between species. Even if questions of evolution were not entirely settled (yet), and the material or visual proof for the process was seen to be lacking, speech seemed to make new ideas of comparative and evolutionary biology *audible.*

CHAPTER 5

Encore

"Mr. Watson, come here. I want to see you." The story is a well-known and oft-reported one. In 1876 Alexander Graham Bell was the first person to use a telephone.[1] Bell's first words across the wire are usually heard as the beginning of a new era: one of fast-paced communication over increasing distances, and one of acoustic technologies that liberated the voice from the confines (or immediate environment) of its producer. More recently, the telephone has served as one of the most important images of the acousmatic voice and of the sound of modernity. Telephones are almost ubiquitous in works of sound studies. For Jacques Derrida it was a "poetico-technical" invention; in Jonathan Sterne's history of sound it represents one of the technologies that disrupt the audiovisual litany; and in Daniel Martin's reflections on the haunting features of the speaking machines of the nineteenth century, the telephone sits among other technologies producing uncanny vocal sounds.[2] Scholars and artists have not only understood the telephone as a harbinger of technological progress and modernity, they have also pointed to its ability to convey presence. In the hands of Jean Cocteau and Francis Poulenc, the telephone became an object that both haunted the human voice and revealed its complexities. As much as the telephone seems to ring in the new, in the context of this book it also seems to spell the end of an era. With the possibility of carrying bodyless voices across wires to other places, the type of materiality of voice that typified nineteenth-century understandings of acoustics, physiognomy, and laryngology seems to come to an end. Acoustic technology—not just the telephone, but the various machines that would follow, like the gramophone and the radio—turns the voice into something eerie again. And while understandings of the voice as a manifestation of the soul could not simply return, the role the gramophone would play in occultism and séances, for example, certainly points to a heightened connection between vocal sounds and the uncanny.[3]

With all this progress and radical change, it is easy to forget the longer history of these technologies. As Jonathan Sterne has pointed out in *The Audible Past*, the stories we are told about these early acoustic technologies are rehearsed narratives that tend to obscure the complexities behind the development, creation, and marketing of products like the telephone and the gramophone. The heroic stories of inventors acting on a flash of inspiration and changing the world celebrate mostly imaginary events. In that context, it bears keeping in mind that Alexander Graham Bell was not only the inventor of the telephone, but also a skilled user of his father's visible alphabet. The telephone, in other words, does not only have a much longer history (and wider genealogy) than its ascription to one person's burst of genius would suggest. It also has a longer cultural history—and even a somewhat dynastic one, which embeds it in practices of imagining and reproducing sound that are ultimately tied to the body and do not require any machinery at all.[4] The principle behind the visible alphabet is amazingly similar to that of the telephone: it attempts to mediatize and transport voice. It also came with an inspiring narrative similar to that of the first phone call, albeit one that has failed to be so widely remembered:

> We, and many others, have seen this method tested in the following way. Mr. Bell sends his two Sons out of the room, and then invites the company to make words in any language, pronounced rightly or wrongly, and sounds of any kind, no matter how absurd or original: for it is the success of this method that whatever the organs of speech can do, the new alphabet can record. Mr. Bell tries each sound himself, until the proposer admits that he has got it: he then writes it down. After a score of such attempts have been recorded, the young gentlemen are recalled, and they forthwith read what is presented to them, reproducing to a nicety, amidst general laughter and astonishment, all the queer Babelisms which a grave party of philologists have strained their muscles to invent. The original symbols, when read sound after sound, would make a Christian fancy himself in the Zoological Gardens.[5]

The visible alphabet, in other words, was a technology that would allow people to hear (or rather, acoustically reconstruct) what had been said in another room with far greater precision than traditional script or shorthand could provide. While the technical execution of this idea in the telephone differs significantly from that of the visible alphabet, there is a cultural continuity between both technologies.

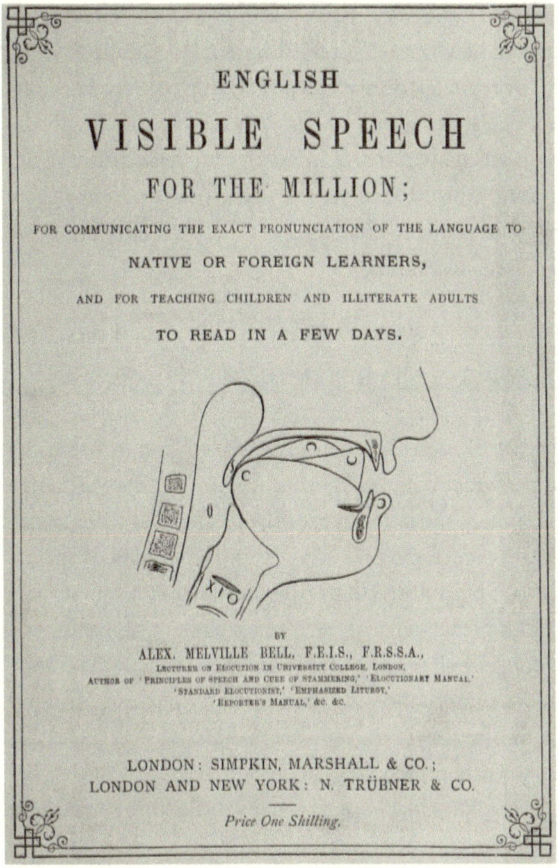

Figure 9. Alexander Melville Bell, *English Visible Speech for the Million: for communicating the exact pronunciation of the language to native or foreign learners, and for teaching children and illiterate adults to read in a few days*, London: Simpkin Marshall, 1868, cover. Courtesy of the John Hay Whitney Medical Library.

Although acoustic technologies have been defined as outside the scope of this book, I do want to take a moment to reflect on the effect such technologies have had on the ongoing imagination and conceptualization of the modern human voice. As the case of the telephone shows, the rise of acoustic technology does not produce a clear or clean break with the more embodied past: ambitions to record preceded the technology; transcription and recording continued to exist alongside each other (and they still do); and the vocabularies of transcription and recording were closely intertwined—as is exemplified by the many manuals for shorthand that presented the method as phonographic and a means to follow speech at its own rhythm.[6] Nevertheless, the rise of acoustic technology, I would argue, produced changes in how voices and vocalic bodies related to each other: rather than erasing the strong connection that had been forged between gendered, aged, ethnicized, and otherwise marked bodies and their concomitant sounds, recordings offered new ways of establishing such naturalized social norms through new means of reproduction and repetition. Moreover, by severing any visible or tangible link between body and voice, recordings brought the ancient connection between voice/speech and humanity under scrutiny again. While acoustic recordings cannot store or revive the voices of the past any more than transcriptions can, they do point to changes in the way the modern voice was performed, imagined, heard, and valued.

A New Era of Disembodiment?

It is important to place the rise of acoustic recordings and—therefore—the rise of the acousmatic voice in the wider context of other scientific and medical discoveries. As Chapter 4 has shown, the latter part of the nineteenth century saw different experts of voice and sound moving away from the organs of speech (and hearing) and toward the brain. Strikingly, the early representations and metaphors to describe the brain leaned heavily on the vocabulary of machinery, electricity, and even the telegraph and telephone. Neurological approaches to the voice not only came to the fore simultaneously with the development of sound technology, the speaking and listening brain was imagined to be a bit like those new, sophisticated machines. This presented a distinct departure from earlier machinery: von Kempelen's speaking machine, and the various automata developed around the same time, were constructed as imitations of the human body, mechanically mimicking its movement and reproducing anatomy. As Adelheid Voskuhl has argued, automata aimed to communicate and arouse sentiment—their cultural work was one of humanizing machinery.[7] When scientists compared the brain to a telephone central, this logic was almost reversed, as it was now the brain that was seen to share the frantic energy of this new technology (they notably also shared a reliance on electricity, with synapses firing away like sparks of energy).[8] Increasingly, machines would be imagined that did not mimic but surpassed human ability—not only in terms of strength or speed, but qualitatively as well. While figures like Edison, Bell, Kratzenstein,[9] Marage,[10] or Watts Hughes[11] were devising tools to reproduce sound with superhuman skill, others like Charles Babbage and Ada Lovelace were developing machines that could "think" beyond human capacity.

Reimagining thought as a matter of connections—as both early computer scientists and neurologists were starting to do in the latter half of the nineteenth century—also led to subtle shifts in understandings of the brain itself. The discovery of Broca's area around 1860 was groundbreaking, but rested on what would soon become an old-fashioned way of grasping the materiality of the brain and its place in the body as a whole. Soon enough, various experts on speech pointed out that, much like the voice was not simply confined to the voice box, speech was not simply produced by a speech center. Rather, voice as well as language were seen as produced through complicated and interconnected physical, psychological, and neurological processes. This resulted in new ways of thinking about vocalization in general, and about speech in particular. Comparative biologists had pointed at differently developed organs of speech, with some characteristics being shared between animals and humans, and others pointing toward the latter's superior evolution. Increasingly, however, it became clear that the brain, too, was an organ of speech and that the capacity to learn how to speak was a matter of neurological activity (and therefore biology, rather than anatomy) as well.

The connection between speech and humanity had a long history, and one that was re-narrated often and purposefully throughout the nineteenth century. The developments in neurology—in conjunction with an increasingly evolutionary understanding of biology—allowed for yet another reformulation of this story. Thinking of the brain as a forger of connections as well as a material part of the human body positioned speech between voice and language in new ways, and made it even more central to a "humanity" that was now part of the animal kingdom. More than the soul, the heart (symbolizing emotion),[12] or sheer Cartesian intellect, speech—that elusive meeting place of body and thought, of voice and language, of uniquely evolved organs and brains—was at once more central to definitions of humanity and more fragile in its exceptionalism. After all, if speech was just a matter of connections, mechanics, and electricity, then who was to say that it would remain exclusive to humans?

The possibility that superhuman voices (as well as superhuman ears) might exist in the near future seems to have been as exhilarating to nineteenth-century individuals as it was frightening. Attempts to reproduce human speech had been made before, with automata, as well as in experiments on dead human larynxes. In that context, it is perhaps no wonder that acoustic technology was sometimes imagined as a way to make the dead speak—with acoustic recordings allowing vocal sounds from those in the beyond to be stored in the here and now. And in some instances, that seems to have been what happened. As John Picker has explored in *Victorian Soundscapes*, the gramophone—seemingly defying death, despite its distinctly earthly origins—would come to play a crucial role not only in the memorialization of famous voices, but also in forms of occultism. *Victorian Soundscapes* shows that vocalities of the late nineteenth century included the voices of deceased poets stored on cylinders as well as numerous departed voices channeled with and through machinery in forms other than acoustic recordings.[13] What it was, exactly, that the machine could bring back is somewhat unclear. The voice had long ceased to be seen as a manifestation of the spirit; but as a vehicle for discourse, emotion, and "personality," it could still acousmatically evoke an absent body and mind.

Unlike Babbage's analytical engines, however, recording devices were not thinking machines: devices that could produce speech would not be invented until well into the twentieth century. Unlike the human, whose body contained a brain with the—still only vaguely understood—capacity not only to understand and invent language but also to transform meaning into sound (and back), the machines were essentially pure body. By the end of the nineteenth century, voice (re)producing machines had ceased to mimic the human body. As anatomical research had given way to a combination of acoustics, comparative biology, and neurology in research on the human voice, the mechanical voice would likewise increasingly depend on physics, acoustics, and flows of information: vibrations would leave physical marks on paper and then wax. Metaphors about other sorts of vocal mimicry dovetailed with this new direction in research and technology: parrots (and so-called "idiots") were believed to speak only in reproduction, after a human voice had made an imprint on their minds. And more literally, sounding like a gramophone would become an insult at the end of the nineteenth century, to be directed at speakers who lacked sense, or whose speech sounded too monotonous to be fully human. This was the case for the unfortunate Mr. Buxton, for example, who spoke at the Cambridge Union in 1895 and who reportedly "argue[d] like a human being but [spoke] as a phonograph would do, when the user of it had been uncertain what to say next."[14]

While we may think of the voice issued from a recording as an acousmatic one, a voice-producing body is still very much implied in the recording. As Kerim Yasar notes in his history of sound reproduction in Japan, acoustic technology subsumed bodily features, with telephone wires stretching out next to vocal cords, and voices settling into the grooves of phonograms.[15] Rather than producing nonhuman voices—as seems to have been a premodern ambition—the nineteenth-century machinery developed around voice was wholly centered on existing human bodies. Recordings were developed to reproduce (and ameliorate) human sound, to facilitate research on the human voice, to gather information on embodied sound rather than mimic its constituent parts . . . and of course, above all, to store the human voice somewhere outside the body (where it would be safe from biological processes of aging and decay). This imagined function of keeping a voice among the living when its owner no longer was, seems to have taken precedence over another important function of recording: the ability to travel beyond its owner's reach.[16] Bell's and Edison's first recordings were focused on great voices that one could expect to be memorable in the (near) future, capturing the voices of Otto von Bismarck, Florence Nightingale, and Queen Victoria.[17]

Use of Recordings to Study Voice in History

This is, generally, how recordings have been used in (historical) studies of the voice: as somehow documenting voices that were performed and could be heard at some earlier time. Katherine Bergeron's use of the *Archives de la parole*, Freya Jarman's careful analyses of the recorded voices of pop stars and divas alike, and even Roland Barthes's reflections on the "grain" of the voice—or lack thereof—on a recording of Fischer-Dieskau are based on an interest in both the recording itself *and* the vocal performance it represents.[18] In most of these cases, it is understood that, first, the voice recorded was one produced specifically for the recording (i.e., it does not document an event that would have taken place otherwise, and it often excludes a reactive audience). Second, these analyses take the cultural context of the recording and the (chronological) gap between the recording and the sound culture in which it can be replayed into account. In doing so, researchers using the act of listening—and a sophisticated analysis of the culturally informed senses of different listeners—have been able to use recordings as witnesses to the sets of expectations that surrounded the acts of singing/vocalizing and of listening, and how both have changed.[19] Works such as those of Nina Eidsheim and Freya Jarman have used recordings to fruitfully examine the links and tensions between vocal sound and its assumed emitting body—laying bare the ways in which gender, sexuality, and ethnicity have been articulated through performative voices.

Moreover, examining the technologies involved in recording voices has proven to be a useful avenue to learn about voice itself. By delving into—rather than trying to block out—practices of mediatization, the recent work of Karin Martensen on the recorded singing voice not only shows the distinctive differences between performances aimed to be live or recorded, but also points out how zooming in on those distinctive qualities of the recorded voice can help us historicize the voice better. And while her attention is squarely on the singing voice—and even more so on the singing voice that could be commercialized through recording—some of her conclusions have a reach far beyond the history of the music industry, which "did not wish to depict vocal 'Reality'" but rather selected and influenced vocal practices.[20] In fact, the many difficulties associated with early recordings as sources (technical differences, but also more culturally rooted methodological ones) have been very useful for me in thinking through a methodological approach toward documents that are not, technically, acoustic recordings but often had the aim of functioning much like them: as ways to reproduce sound, and particularly the human voice.

Although the primary materials used for this book are largely visual, and consist of or closely resemble text, the consideration one needs to bring to their production, invention, technical characteristics, and use is very similar to that of acoustic recording, and requires an interest in its various mediatizations.[21] Moreover, as the thorough medical investigations of the nineteenth century show, voice itself can be considered a technology as well, not only because particular kinds of vocalization require technique, but because even the most commonplace practices of speaking, singing, and listening are informed by (shared) cultural knowledge, by years of training, and by (un)written norms and expectations. Recognizing that vocalization is a technology—albeit a deeply human and embodied one, with a remarkably long history—goes against a well-established tradition of dualism. Walter Ong's notions on the technology of writing as one that "structures thought," and punctuates societal rhythms moving from orality, to literacy, and to new oralities back again, are somewhat more complicated but have nevertheless largely been read as clear expressions of such dualistic distinctions.[22] The distinction between writing as technology and speech as a natural human ability not only echoes the kind of divide Jonathan Sterne called attention to with his critique of the audiovisual litany, but also constitutes part of the anthropocentric litany that has recently been called into question from a posthumanist perspective.

The development of complicated machinery that mimics or supports the work of the senses (like vocaloids or cochlear implants) seems to be turning us increasingly into the cyborgs Donna Haraway already envisioned in 1985.[23] Taking into account the cultural work of developing sonic skills as well as vocal training, and questioning the divide between the technology of writing and the ability to speak, shows that we have, in fact, been cyborgs all along. Moreover, we have been sharing our cyborg qualities with both animals and machines. Despite foregrounding the ability to speak as a central and defining characteristic of humanity, nineteenth-century researchers (like eighteenth-century builders of automata before them, and inventors of various acoustic devices after them), have made significant efforts to isolate the voice from the human body, and give it a life of its own. Despite the steep rise in acoustic technology, however (we are now almost constantly surrounded by recorded and synthesized voices as well as machines that take our calls without human intervention),[24] we remain attached, physically and emotionally, to our own and others' voices. This is, perhaps, why we get so annoyed with speaking machines like Alexa or our GPS navigators when they don't understand us: there is still an imaginary human hiding behind those synthetic voices, at which we direct our emotions.

Similarly, the acquisition and systematization of knowledge about the voice keeps circling back to human-embodied experiences of sound. The consistent push toward visualization and detached objectivity did not result in a removal of the expert's body from the field of vocal knowledge. On the contrary, intimate familiarity with one's own sensations and vocal experiences became even more crucial to fields such as laryngology when visualization of the throat became widely available: connecting the produced images to experiences of sound, proprioception even emerged as a significant experience in trying to understand and interpret these images. The diagnosis of vocal pathology would never become the exclusive terrain of observing the throat, but would come to rest on an interconnected system of hearing pathology in vocal sound, seeing the voice at work, and interpreting both of those pieces of information through a skillful comparison of feeling one's own healthy/normal voice against a pathological or abnormal one.

Perhaps it may need to be pointed out that the way in which current researchers in voice studies and the history of the voice go about their work is not entirely dissimilar to that. Although large parts of the material used are visual or textual in kind, the other senses necessarily come into play as the researcher's body and ingrained sonic skills unavoidably interact with the historical bodies, cultures, and performances under scrutiny. If used merely intuitively, the researcher's body and sensory skills may appear as a problem. As pointed out above, not possessing the "period ear" will make us hear differently, or even different things, than contemporary listeners would have. I would argue, however, that there are ways to mobilize the body that move beyond simplistic notions of empathy across the ages and can therefore be helpful in learning to recognize historical difference.[25] Thinking about historical as well as current actors as cyborgs—rather than members of a seemingly universal humankind—can serve as a first step toward that. Examining the prescribed/described physical practices of the nineteenth century as the cyborg that I am today (not simply as a fellow human with a similar body, but as a creature with a particular set of ingrained skills and expectations), and engaging in practices that have become slightly changed or entirely unfamiliar, show sites of friction and moments of difference. Like encountering a piece of vocabulary no longer in use, feeling my throat rebel against an exercise I have come to see as "unnatural" or even impossible flags moments of historical difference—or, at the very least, it can jolt me into reexamining any assumptions that I've brought to my reading.[26]

Projections of Norms and Values

What all this shows, once again, is that recordings do not simply retain acoustic events or performances of the past, but rather that recording machinery and its products are part of a large cultural field in which bodies, sounds, electricity, material objects, and immaterial experiences and ideas collide. The fact that we no longer consider early recordings as hi-fi renditions of actual events or sounds in the past is therefore not just a matter of technological progress (i.e., that our machines now produce a better quality of sound) and not even only a matter of changing expectations (i.e., how to nineteenth-century ears the phonograph sounded clear because it far surpassed anything that had been available before). Rather, our understanding of what a recording should convey has changed over time, as have the sonic skills that allow us to hear a recording as something other than a collection of sounds put together in a studio. The recording is, therefore, not so much a document of a particular acoustic event, but rather a foil onto which norms, expectations, and values surrounding the (singing) voice have been projected—and onto which the contemporary researcher may have trouble not projecting her own.

In the introduction to this book, I referred to Nicholas Cook's definition of the recording as a methodologically helpful one, allowing for an understanding of the cultural work done, diegetically, by both the act of recording and the practice of listening to recordings.[27] It bears keeping in mind, however, that Cook's observation, too, is historically situated and owes much to the practical reality of recording in the late twentieth/early twenty-first century. While I would argue that this approach to any understanding of a recording is a helpful one, examining the kind of performance that is constructed through the audience's act of listening, and the particular practices of recording sound toward an acoustic document that will help a listener to do so, is still of importance. Cook's observation reminds us, in general terms, that although recording devices are designed to reproduce an embodied sound, recordings produce more than they reproduce performances, and they do so culturally. In other words: norms, values, and esthetics of the voice in a general sense inform the status, reception, and indeed production of the recording . . . and they can do so on a very practical and material level, influencing the shape, distance, and technology of the microphone; directing the position of singers and accompanists; distributing the roles of expert between sound engineer, producer, and musician; and indeed initiating changes in vocal technique—and thus the singer's body itself.

Early acoustic recordings in the late nineteenth and early twentieth centuries constitute an additional—and different—set of documents to study the tail end of the vocal practices of the long nineteenth century. They are not necessarily less normative in kind than the prescriptive and descriptive materials used here. Yes, these machines recorded actual rather than imagined or remembered practices, but they first recorded only those practices considered exemplary enough to warrant (expensive) recording, and, second, they depended heavily on the listener to gain any sensorial meaning. In other words, even if a machine is involved, the voice remains an intersubjective phenomenon. Researching the different moments and methods of its mediatization is difficult and necessary work to come to an understanding of vocal practice, but should not preclude or overshadow inquiries into the voice's embodied particularity: that human organs produce and receive it simultaneously and that, therefore, hearing and vocalizing are intimately connected.[28] Not only do we hear ourselves when we speak, when we hear another's voice we rely on the experiences of our own vocal organs to interpret those sounds as signs of identity, health, and social status (through either similarity or difference).

What acoustic recordings offer, then, is—much like recordings on paper—an invitation to imagine a vocalic body engaged in a performance of beauty, health, or power. By recording the voices of Bismarck, Queen Victoria, Sarah Bernhardt, or Robert Browne, one could imagine that these great figures could somehow enter one's living room. This is also how phonograph companies (and later the radio) advertised their wares: the quality of the recording was not necessarily expressed in how well it represented actual performance, but rather through how well a recording would evoke the presence of a singer like Caruso in one's living room.[29] Even as these machines eliminated the need for physical proximity to have an experience of a famous voice (thereby eventually democratizing access to both opera and politics), they also helped to cultivate a culture in which voice, presence, and the body were intimately interlinked. The celebrity culture that would develop around nineteenth-century operatic divas bears witness to this phenomenon. As James Davies has shown, the diva was born out of modernity.[30] The availability of recordings of their singing would continue to play an important role in the visceral experience of their voices—especially by fans—in later periods.[31]

Although acoustic recordings therefore mark the end of an era (and the beginning of a new one), they also bear witness to the changes that were taking place in embodied voices throughout the nineteenth century. As Cook's analyses of recordings suggest, these documents and technologies produce voices rather than reproducing them. In doing so, they also produce bodies—at least in the listeners' and the wider cultural imagination. The voice itself, as Steven Connor has highlighted, alerts us to a being that is given voice; it gives substance to the self.[32] After the nineteenth century, in which the clothing of the voice in cartilage, sinews, and muscle was so explicit, the body that is created by the recorded voice is more Frankenstein's monster than Coppélia—the result of the scientist's work rather than the artist's or artisan's. That is not to say that fascination with the statue brought to life, and the acousmatic voice of spirituality that attends it, simply disappeared, but that the vocal uncanny and the acousmatic voice changed along with the embodied voice throughout the nineteenth century. Moreover, as the voice—live or recorded—entered the twentieth century, the borders between the anchored and acousmatic voice became increasingly blurred. Increasing familiarity with recording as well as changing vocal techniques in the light of recording and amplification have an effect on the human voice and its meanings, much as earlier scientific discoveries in biology, pathology, and acoustics did.

This blurred line between seemingly immediate and recorded vocalizations explains, perhaps, why investigations into the twentieth-century voice have focused on electrified voices—that is, voices that are mediatized but neither necessarily acousmatic nor entirely tethered to the body. In a sense, my reasons to focus in this book on the medical and educational technologies surrounding the voice are not so different: there too, the blurred line between the materiality and intangibility of the voice is at stake, constantly reimagined by authors who are coming to see both the human body and its place in nature in a new light. It is in exploring the porous border between the anchored, physical voice and the untethered, acousmatic one that new technologies seem to emerge, and it is also where changes regarding the cultural and political meaning of the human voice can be observed. In the beginning of the twentieth century, when recording equipment starts to occupy and police the border between the voice-in-the-body and the acousmatic voice, this rings in a radical remaking of the voice for modern times, but one that owes its existence to a long tradition of engaging with that border—and of crossing it—both practically and in ambitious imaginings.

While the voices that are produced by early twentieth-century technology are called to life by Frankenstein rather than Dr. Coppelius, guided by a belief in the innovation of science and electricity more than the cumulative knowledge of craft and skill passed on intergenerationally, they are still connected to these longer traditions of craft. Much like the modern nineteenth-century singer owes many of her embodied practices and vocal features to the repeated and inherited technique of the castrato,[33] so too do the phonograph, radio, and telephone voices of the twentieth century owe their existence to a long history of imagined and real speaking machines and living dolls that populated the salons and tales of the eighteenth century, much in the way that Hoffmann would come to occupy the nineteenth-century operatic stage. The technology, and the philosophical ambition, of such speaking dolls is very different from that of the gramophone, of course, but like so many of the practices and documents described in this book—both engage in a performance of representation of the human voice, and both do so by means of alternative embodiments. Pulls and levers, electrical charges, or lines on a page are made to stand in for an absent human body, while the acoustic experience provided by all these media aims to bring the inanimate stand-in body to life, like a ventriloquist dummy without a puppeteer.[34] Whether it is the gramophone delivering Nellie Melba to your living room, or Bell's visible speech transporting a speaker of some unknown, faraway language to a hall full of stunned professors, these documents rely on the strong cultural association between voice and presence to breathe physicality into documented or materialized sound.

Conclusion

The history of sound reproduction is, as James Davies has argued, "a tale beholden to the history of cultural techniques," and the rise of acoustic technology did not lead to "the voice's severance or rupture from the body so-called."[35] In his examination of the phonographic recording of Jean de Reszke's last performance, Davies shows how the phonograph, in its late nineteenth-century iteration at least, had no ambition to sever voice and body but rather aimed to freeze the atmosphere of this last performance. Doing so, Davies points out, "crystallized" de Reszke's particular vocal technique and the effect it had on his audience, "less for playback than as mystic keepsakes for archival storage."[36] If later, commercialized recordings would come to be used as representations of sounds and their producers, conjuring their singing bodies out of thin air, the very early archival recordings were devoted "to a kind of disciplining of the air: in the abstraction, subjugation and imaging of the living atmosphere itself."[37] As Davies points out, this ambition to discipline the air was shared by both archivists and singers, whose vocal techniques at that time were increasingly concerned with the development and mastery of breathing techniques.[38]

De Reszke, unlike other singing masters who have been more visible in this book, like Garcia, Seiler, and Marchesi, did not rely on laryngoscopic examinations or imaginations of the throat. He developed a more intuitive, breathing-based technique (one still known to many singers now, who continue to sing into the mask).[39] And yet, like Garcia, Seiler, and many others of the physiological school of singing, proponents of this "school of the nose," too, relied on imaging technology. Recording technology, it appears, not only did not sever the voice from the body, it also did not discourage further investigation into the visual representation of (vocal) sound. The technological tales "beholden to the history of cultural technique" that arose in both the nineteenth and the twentieth centuries are therefore multiple and interwoven. The documentary fervor of early statisticians, gathering numbers on the prevalence of stammering, lives on in the librarian keen to archive de Reszke's vocal technique, while the search to see the voice by use of the laryngoscope is mirrored by an equally feverish hunt for means to preserve vocal sounds and freeze them in time much later.

Moreover, the cultural techniques at play here extend far beyond the world of music. As Davies notes, "Technologies . . . are mere crudescences of consolidated cultural behaviours, vocal technologies are the mere accoutrements of wider techniques of projecting the voice."[40] These techniques—not only of projecting the voice, but also of managing, manipulating, and adapting the voice to changing circumstance or to suit a profession—gave rise to a range of vocal technologies at play far away from the operatic stage. The identification of the pre-phonograph era as a time of "sovereign vocalism"[41] was due to grandiose performances at the pulpit, on the political platform, and on the theatrical stage as much as it was to operatic performance. As the previous chapter has shown, the development of vocal technology during the long nineteenth century was not confined to even that wide variety of public and staged voices either. The so-called normal voice, lodged in unexceptional throats, was perhaps even more central to the endeavors of many nonmusical observers, and became a site for the development of vocal technologies too. The recording technology that would eventually grow out of *all* these techniques, consolidating a very wide range of cultural behaviors, can hardly be said to have suddenly ended an era of heightened attention to embodied vocality. It does, however, represent a radical change in the ways vocal technologies are documented—and therefore also in the methodologies needed to study them.

Conclusion

Generationality, Humanity, and Listening Against the Grain

When I was doing the research for this book, poring over medical treatises, singing exercises, elocution manuals, and stuttering songs, the greater arc of the story that took shape in my head was one of materiality, of tangibility. All these efforts to catch the voice at work: seeing inside the larynx, visualizing its sounds, pointed to a gradually formed but profoundly modern characteristic of the nineteenth-century voice: it was utterly embodied and material. Whereas acousmatic voices were embraced (if sometimes ambiguously so) in periods before and after, the sourceless voice was deeply distrusted in the nineteenth century, as experts were committed to fully understanding the voice as something produced by the vocal organs. This evolution can be partly explained by the growing specialization of experts—who would be phonologists or laryngologists, rather than natural philosophers—but I would argue that their modes of seeing, hearing, and specializing still say something about how the voice was conceptualized in the nineteenth century, and it was made more viscerally material in the age of biology and evolution than it had been in that of anatomy. The chapters of this book are organized loosely around this logic: they show how the rise of the embodied and material voice was facilitated by experts of different stripes. They examine how the embodied voice took shape in (and in turn gave shape to) different bodies, or how material bodies shaped not only voices but practices of vocal knowledge as well. They trace how knowledge about the voice changed in conjunction with changing conceptions of the body when the brain became more integrated with understandings of the embodied self. Finally, they consider how acoustic recording practices heralded the return of the acousmatic voice. The long nineteenth century, so it seems, can be read as the age of vocal materiality—a relatively brief and exceptional period in time in which the voice's eerie qualities were distrusted more than they were accepted, and were therefore pushed to the side.

However, such an image may give the impression not only of exaggerated exceptionalism, but also of stasis. As if, in the nineteenth century, the slow changes that had been taking place throughout the preceding centuries in how "we" understand "our" voice were interrupted, and a period of leaps in scientific observation brought these slower evolutions and perambulations in vocal knowledge temporarily to a halt. It is hard not to understand the nineteenth-century voice as defined by the two major inventions that seem to dictate the rhythm of the rise of the material, embodied voice, and the eventual return of the acousmatic one. The introduction of the laryngoscope and the discovery of the speech center, both around 1860, provide a strong chronological anchor point for this study. They vividly show the tight interaction between different fields of knowledge (as well as boundary work) in the development of vocal expertise; they provide insights into the ways in which voices could be imagined in the nineteenth century (and crucially how these images differed from those of their predecessors); and they show how expertise of the voice was established and unequally (re)distributed in this age of science, medicine, and pedagogy. Importantly, however, these two moments of invention also overshadow other changes in the professional vocabulary about the voice. While the appearance of laryngoscopy and Broca's area are highly visible and traceable moments of change—one can write genealogies of these ideas with reasonable ease and precision—they may not account for all changes taking place in a field of knowledge that, after all, was by no means an established or clearly defined one: access to knowledge about the larynx and speech center was not immediately available to all, nor was it necessarily sought by everyone.

So, by way of conclusion, I want to reimagine the vocabularies, images, and performances introduced in the preceding chapters as signals of other changes, as part of a longer, slowly meandering narrative of thinking about and understanding the voice. Doing so takes us right back to Plato and Aristotle, the early experts to whom the close connection between voice/speech and humanity was so often ascribed. For while both voice and humanity are historically specific thoughts, connected to meanings, practices, and discourses that have changed over time, their connection has proven to be surprisingly resilient. We continue to understand voice as a manifestation of subjectivity, an avatar of a unique and individual, or soulful and belonging, human being.[1] This connection between voice, subject, and humanity, I would argue, became deeply implicated in the imagination of the material, embodied voice in the nineteenth century—but cannot simply be reduced to yet another story of modern individuality and biology. Rather, we first need to listen back to a different rhythm of change, and another understanding of humans' place in history: one dictated by generationality rather than individuality. If we are to distinguish an age of vocal materiality, it is among these generations of bodies and experts that such an age can be found, and it must necessarily be one for whom the human is at stake.

Even though modern notions of innovation, invention, and individual achievement are central to the most obvious stories of the history of vocal knowledge in this period, they are constantly intertwined with companion stories of family resemblance, inherited skill, and intergenerational exchange. This is most visible (as noted in Chapter 2) in the field of canto, where long dynastic traditions of artistic skill and fame existed and were kept alive throughout the nineteenth century, and even through to today. Whereas the famous castratos of yore were no longer teaching at this time, their memory was kept alive by their pupils' pupils, and some of their methods survived the rise of a more scientific pedagogy by being integrated into the breathing and singing exercises of the new age: the old masters were thus newly credited with their skills in preserving and cultivating a naturally beautiful voice. This intergenerational inheritance scheme of skill and beauty was also at play in the invention of the laryngoscope. With his long life, which spanned the whole nineteenth century, Manuel Garcia bridged the gap between the old masters and the age of recording, both of which were heavily reliant on the ear, with a virtuoso display of visualization. The success of his method was acted out as a multigenerational one, including the reputation of his father, who was a famous canto teacher in his own right, but above all relying on the fame of his pupils—his sisters (Pauline Viardot and Maria Malibran) and young women like Mathilde Marchesi who would carry his knowledge into the twentieth century.

And so the age of vocal materiality and its innovative heroes look somewhat different than they may first appear. The laryngoscope and speech center were often represented as singular achievements, performed by great men whose bravery, intrepid sense of discovery, and individual intellect would later be celebrated in tones reminiscent of those used to describe the explorers of new continents. But such stories were always underpinned by narratives of knowledge and skills with much deeper roots, and a much wider application. Garcia and Broca were, within their own small circles, celebrated as geniuses, but as geniuses preceded by "fathers" of different stripes, surrounded by supporting wives, sisters, and children, and polished through interactions with a growing class of experts across generations. The collective of experts also, curiously, grew backward. That is, great masters and classical philosophers were recuperated as members of the new vocal disciplines in order to give them more legitimacy in the present and a firmer footing in the past. The self-declared fathers of innovation and invention were all too happy to establish connections with well-respected grandfathers of their field—be they famous castratos or ancient Greek philosophers.

Despite nineteenth-century experts' great willingness to mention, cite, and engage with classical authors' interpretations of vocal physiology and practice, their interpretations of the voice's role as defining the speaker ("speak, so that I may know you") were quite different from those expressed by Aristotle or Quintilian. More than a century of first exploratory and then systematic study of the larynx had profoundly changed the ways in which the voice, its relation to the body, and its material presence in the world were understood. In the age of vocal materiality, the human voice had become both more natural—that is, rooted in biology, connected to instinct—and more subject to cultivation and efforts to cure, improve, or strengthen it. The result of these somewhat contradictory processes was a renewed need to assert and demonstrate the humanity of the voice. For if the voice was natural, could it not also be produced by animal throats? And if it was the result of artifice and cultivation, maybe machines could acquire vocal abilities as well, as seemed to be the case with inventions such as the vowel machine or the phonograph. At the heart of these questions were not so much the obvious points of distinction between human, beast, and machine, but rather the places where the categories seemed to overlap. Even when firmly lodged in the throat, something about the voice remained uncanny, and its uncertain capacity to articulate humanity led to cultural, political, and scientific anxiety.

Despite nineteenth-century vocal experts' love of the Aristotelian dichotomy of animals without and humans with the gift of speech, the distinctions they drew in their own work were not nearly as neat. For modern laryngologists and voice educators, humanity was an unclear, ambiguous, and above all fragile category. Joanna Bourke's *What It Means to Be Human* sketches how social and cultural imaginations effected changes in the designation of humanity, and the exclusion of numerous groups such as women, "savages," disabled persons, and so on.[2] Citing Kafka's scholarly speaking simian in *Rote Peter*, she draws our attention to the cultural imperative of socially sanctioned language for one to be recognized as human.[3] More importantly, however, her history of the animalistic audition of women, children, impeded speech, and languages outside of the West points to the changing anxieties surrounding the limits of the "community of men."[4] The continuous reliance on Aristotle, the lengthy scholarly essays on conscious intelligence, the frantic attempts to wean children off animalistic behavior all seem to be thinly veiled articulations of an ultimately very emotional practice of defending humanity as an exclusive and politically performative category. Bourke asks, in her history of humanity, "might the insistence on the importance of language as a marker of status, even full personhood, be an act of violence?"[5]

However, as the parade of speaking parrots and parroting children studied by comparative biologists showed (see Chapter 4), full personhood was connected to speech in ways that extended far beyond language. In the age of vocal materiality, speech was an explicitly embodied activity, lodged in anatomical features and expressive of emotion and passion as much as of reason and intelligence. The insistence on its centrality in defining the human was therefore not only an act of cultural and political violence, it was also articulated through physical violence. Making speakers human was achieved by cutting stammering tongues, by institutionalizing the dysfluent and the speechless, and by the practice of various methods of laborious training. The material manipulation of dehumanized bodies thereby reflected the changing ways in which species were distinguished. In a Linnaean world of obvious hierarchies and dichotomies, only the surgical knife could bring about real change, whereas in a Darwinian world of kinship and evolution, only the constant practice of speech could bring about (vocal) humanity.[6] As Joshua St. Pierre notes in *Cheap Talk*, this discourse continues to exclude a range of different modes of speech: "what is common to disabled speech, in its many expressions, is an assumed unintelligibility that results in an unevenly distributed failure to be recognized as a rational and human (political) agent."[7] Meanwhile, in the nineteenth century, the seat of speech—in both parrots and humans—shifted from the tongue to the brain, and so parrotness and humanness came to be located in the brain as well. The distinction between these two brains therefore lay not so much in the capacity of one of them to acquire language, but rather in their ability to produce subjectivity through speech.

This insistence on the humanity of the ability to vocalize subjectivity as material presence also resulted, paradoxically, in attempts to hide the imperfect body that produced such speech. Overcoming imperfections, producing fluent speech or a smooth bel canto voice, was ultimately meant to express humanity without the obstacle of its difficult and disruptive physicality. The bel canto technique encouraged singers to smooth over the break between vocal registers, in other words to hide the work of the larynx by relentlessly focusing on and training that organ. Stammerers were to hide the hard work an unruly tongue necessitated by working even harder, much as deaf pupils were taught to engage different senses and parts of the body to learn to look as if their lips had always known how to move naturally. If the nineteenth century is the age of vocal materiality, that is largely because so much energy was poured into the creation of a fiction—not so much of disembodiment, but rather of miraculously fit, proper, faultless bodies—so faultless, indeed, that their presence seems to melt away before the humanity and subjectivity they housed. The parrot, needless to say, never managed this very human sleight of hand: exercising its voice more only drew more attention to the flaws that made its voice so parrot-like.

The modern human voice, however, seems to have been remarkably successful at performing its trick of embodied disembodiment. By the twentieth century, the well-trained, smoothed, fluent voice could still fool listeners into believing that it hardly issued from a body at all, instead giving the soul an immediate, seemingly unmediated presence. This, at least, is what Roland Barthes claimed to hear when listening to Dietrich Fischer-Dieskau. Notably, he heard the German baritone as an acousmatic voice, on a long-playing record whose sound quality presented a remarkable improvement from the phonograms that preceded it. The well-trained voice, with perfectly managed breath—according to Barthes—"never transcends culture."[8] It is pure pheno-song,[9] bypassing the body. This bodyless performance of culture is compared to the voice of Panzéra (who was not recorded on long-playing records, as Barthes notes) and of an imagined Russian bass. In these less smooth, gravelly voices, Barthes heard: "That which is past (or previous to) the meaning of the words, of their form . . . and even of the style of performance: something which is directly the singer's body, brought by one and the same movement to your ear from the depths of the body's cavities, the muscles, the membranes, the cartilage . . . as if a single skin lined the performer's flesh and the music he sings."[10] For all the snobbery inherent in Barthes's account (the popularity of Fischer-Dieskau is dismissed as petit bourgeois, the perfection in recordings as a sign of a boring longing for predictability), we can read his account in *The Grain of the Voice* as a reflection on what he—with educated mid-twentieth-century ears—heard. Listening out for the "muscles, the membranes, the cartilage," he could not discern them in Fischer-Dieskau's voice—which was, after all, like any other, issued from a body. In Fischer-Dieskau, who had a thorough but (in World War II) interrupted formal training, the modern ideal, formulated and put into practice in the nineteenth century, was attained: he presented a perfected nature by cultivating his voice, training the body into making itself invisible. Or it has, at least, when directed at ears with training in a very similar cultural arena. Barthes's reflection, his project of "shifting the fringe of contact between music and language" rather than fruitlessly trying to write about the "ineffable," leads him to what has also been called the musical flaw,[11] to moments of rupture where the body-in-song becomes discernible. For Barthes, "the grain is the body in the singing voice, in the writing hand, in the performing limbs."[12] My project here is, in the end, perhaps not so different. This book, too, has chased after the "grain," only to discover that the perfect pheno-song hides one too. Whereas Barthes addresses singing teachers' obsession with breath as a myth, as a trick to bypass the body, this book has attempted to open the magician's toolbox and lay bare the mechanisms, the hard work—physical as well as cultural—that made the modern voice, and made it sound like pheno-song. If the grain is the body in the singing voice, then this book argues that it should include a multitude of bodies, giving equal billing to the stammering tongue and the managed breath; to the perfected soprano trill and the leaps of the pubescent boy's "breaking" voice; to the silver-tongued politician and the cackling fishwife.

In that sense, the history of the voice is not a niche pursuit, relevant only to those who deal with professional singers or great orators. While we may, usefully, question Aristotle's equation of speech with humanity, the human voice is an important aspect of the human experience and—in many ways—an actor for historical change. Not every voice has launched fleets, or had the kind of wide political or cultural impact of a Gladstone, a Lamartine, a Pankhurst, Patti, or Caruso, but decisions are made, families forged and disbanded, business deals closed, to the tune of a variety of human voices. As Freya Jarman has noted, the voice (and therefore, all these voices) is "naturalized rather than natural as such," but nonetheless "complicit in the theatrics of gender" and other categories of identity and identification.[13] This book, therefore, hopes to offer opportunities to include voices (the spoken word, but also songs, hums, whispers, and stutters) in analyses of daily social and cultural practice; of the performance of gender and race; of the experience of childhood; of narratives of sexuality and depictions of parenthood; of the stirring of emotions such as bravery, patriotism, and anxiety; and of the rituals of death. Recognizing that voices are never given, but always practiced, performed, and intersubjectively experienced, has consequences in particular for the modern period. That it presented itself as the age of print, and was peopled by witnesses who understood the voice as natural, has, perhaps, encouraged us to think of it as a period in which metaphorical voices (the vote, the voice of the people) mattered more than actual vocal practice. The many stammerers, singers, speakers, and shouters who bought and populated the pages of the stream of publications on vocal practice and health in the nineteenth century would probably disagree.

NOTES

INTRODUCTION

Epigraph: Marc Colombat de l'Isère, *A Treatise upon the Diseases and Hygiene of the Organs of the Voice,* Boston: Redding & Company, 1857, 15.

1. Gina Bloom, *Voice in Motion: Staging Gender, Shaping Sound in Early Modern England,* Philadelphia: University of Pennsylvania Press, 2007, 66–110.

2. Theodore E. Schmauk, *The Voice in Speech and Song: A View of the Human Voice for Speakers and Singers and All Who Love the Arts of Speech and Song,* New York: John B. Alden, 1890, 113.

3. Ivan Kreilkamp, *Voice and the Victorian Storyteller,* Cambridge, UK: Cambridge University Press, 2005, 6: "Speech, increasingly, becomes the sign of the human and the humane."

4. John Heath, *The Talking Greeks: Speech, Animals, and the Other in Homer, Aeschylus, and Plato,* Cambridge, UK: Cambridge University Press, 2005.

5. Pierquin de Gembloux, *Idiomologie des animaux ou recherches historiques, anatomiques, physiologiques, philologiques, et glossologiques sur le langage des bêtes,* Paris: à la tour de Babel, 1844, 3.

6. Colombat de l'Isère, *A Treatise,* 20.

7. "It is," according to Jacques Derrida, "implied in the very structure of speech that the speaker *hears himself,*" yet only in an act of "pure auto-affection"—not through the intervention of an external mechanism. Jacques Derrida, "The Voice That Keeps Silence," in Jonathan Sterne (ed.), *The Sound Studies Reader,* London: Routledge, 495–504, 497.

8. See, e.g., Cecilia Bartoli's remarks on "the magic of the natural voice" and "listening to your instrument" (http://www.theguardian.com/music/2005/oct/14/classicalmusicandopera).

9. Although the nineteenth-century voice's connection to authority is as strong as the eighteenth century's, its connection to physicality is quite different: whereas—as Gina Bloom notes—the instability of the body plays an important role in the imagination and meaning of the early modern voice, in the nineteenth century the instability of the voice itself is at issue, while the body is imagined as its stable material seat. Bloom, *Voice in Motion,* 2–3.

10. Adelheid Voskuhl, *Androids in the Enlightenment: Mechanics, Artisans, and Cultures of the Self,* Chicago: University of Chicago Press, 2013.

11. E.g., Elizabeth Stevens, *Anatomy as Spectacle: Public Exhibitions of the Body from 1700 to the Present,* Liverpool: Liverpool University Press, 2011.

12. The voice only exists *during* speech, as has been noted, e.g., by Mladen Dolar and Arlette Farge. It is a "vanishing mediator." Mladen Dolar, *A Voice and Nothing More,* Cambridge, MA: MIT Press, 2006, 15. Arlette Farge, *Essai pour une histoire des voix au dix-huitième siècle,* Paris: Bayard, 2009, 21.

13. The history of the body has therefore been written as one of slow and subtle change, in which the process of construction that is at issue is always hidden beneath a cloak of seemingly given natural features and shapes. Theories of the history of the body insist on the need for constant repetition and reiteration (or, in Foucauldian terms, discipline) for the historically appropriate body to be kept intact, but also for ideals and practices to change (Willemijn Ruberg, *History of the Body. Theory and History,* London: Bloomsbury Academic, 2020). Although the cultivation of the voice entails similar amounts of cultural and physical work, this work was far less hidden to the contemporary listener.

14. Madeleine Hurd notes, for example, that working class speakers attempted (successfully) to adopt bourgeois modes of dress, behavior and speech to gain access to the public sphere in the nineteenth century. Madeleine Hurd, "Class, Masculinity, Manners and Mores. Public Space and Public Sphere in Nineteenth-Century Europe," *Social Science History,* 24:1 (2000), 75–101.

15. Nineteenth-century science was a space of polemics throughout Europe, and the practitioners of these "new" disciplines shared their style with their colleagues in other fields. See, e.g., Raf De Bont, "'Writing in Letters of Blood': Manners in Scientific Dispute in Nineteenth-Century Britain and the German Lands," *History of Science,* 51:3 (2013), 309–335.

16. As Thomas Gieryn has noted, "Boundary-work would be expected in settings where tacit assumptions about the contents of science are forced to become explicit: where credibility is contested." In the constant confrontation between different emerging disciplines concerned with the voice, that is certainly the case. Thomas F. Gieryn, *Cultural Boundaries of Science: Credibility on the Line,* Chicago: University of Chicago Press, 1999, 24.

17. "It sometimes, indeed, happens that a massive intellectual and physical presence, is accompanied by a thin, shrill or piping voice. But that this is not what we ordinarily expect, is amply shown by our feeling of disappointment and our sense of incongruity of the case. Dr. Johnson growled, Carlyle thundered, Coleridge piped. The breadth of chest and strongly developed pugilistic qualities of John Angell James and his powerfully supporting physical accessories in speaking, would lead us to expect a voice to correspond to its immediate source and environment." Theodore E. Schmauk, *The Voice in Speech and Song,* 109.

18. See, e.g., Anne Heyer and Anne Petterson, "Learning from the Outside: Parliament's Response to Public meetings in Germany and the Netherlands, 1870–1914," *Parliaments, Estates and Representation,* 44:1 (2024), 64–79; Martin Hewitt, *Making Social Knowledge in the Victorian City: The Visiting Mode in Manchester, 1832–1914,* London: Routledge, 2020.

19. Denyse Rockey, *Speech Disorder in Nineteenth-Century Britain,* London: Croom Helm, 1980, 217.

20. Paul Broca's discoveries on aphasia, and on the connection between linguistic capabilities and the integrity of the brain, are usually seen as the advent of neurolinguistics. Broca presented them in Paris in 1861, at the *Société d'Anthropologie.* L. S. Jacyna, "The 1874 Aphasia Debate in the *Berliner Gesellschaft für Anthropologie,*" *Brain and Language,* 69 (1999), 5–15. Paul Broca, "Du siège de la faculté du langage articulé dans l'hémisphère gauche du cerveau," *Bulletins de la Société d'anthropologie,* 6 (1865), 1–18.

21. The introduction of recordings would, according to Dmitri Zakharine, have important consequences for the voice itself as well: human voices would be gradually replaced by "an abstractly understood 'vocality,'" consisting partly of the acousmatic voices produced by radio, telephone, and other machinery. Dmitri Zakharine, "Preface: Electrified Voices. Medial, Socio-Historical and Cultural Aspects of Voice Transfer," in Dmitri Zakharine and Nils Meise (eds.), *Electrified Voices:*

Medial, Socio-Historical and Cultural Aspects of Voice Transfer, Göttingen: V&Runpress, 2013, 12–29, 12.

22. On the association between audio(visual) media and the uncanny, see Jeffrey Sconce, *Haunted Media: Electronic Presence from Telegraphy to Television*, Durham, NC: Duke University Press, 2000.

23. Victor Records' early twentieth-century advertisements for recordings of Enrico Caruso claimed that one could not hear the difference between their recordings and the "real" thing: "Both are Caruso," according to Victor. Jonathan Sterne, *The Audible Past: Cultural Origins of Sound Reproduction*, Durham, NC: Duke University Press, 2003, 215–216.

24. This also affected the meaning, and the sound, of voices—especially in the public sphere. See, e.g., Kate Lacey, "Speaking Up and Listening Out: Media Technologies and the Re-Sounding of the Public Sphere" in *Electrified Voices*, 123–136.

25. According to Mladen Dolar, it is "a bodily missile which has detached itself from its source, emancipated itself, yet remains corporeal." Dolar, *A Voice and Nothing More*, 73.

26. John Potter, *Vocal Authority: Singing Style and Ideology*, Cambridge, UK: Cambridge University Press, 2006.

27. John Potter and Neil Sorrell, *A History of Singing*, Cambridge, UK: Cambridge University Press, 2012.

28. Martha Feldman, Emily Wilbourne, Steven Rings, Brian Kane, James Q. Davies, "Why Voice Now?" *Journal of the American Musicological Society*, 68:3 (2015), 653–685.

29. James Q. Davies, *Romantic Anatomies of Performance*, Berkeley: University of California Press, 2014, 66–92; idem, "Pneumotypes: Jean de Reszke's High Pianissimos and the Occult Sciences of Breathing," in David Trippett and Benjamin Walton (eds.), *Nineteenth-Century Opera and the Scientific Imagination*, Cambridge: Cambridge University Press, 2019, 21–43.

30. Arlette Farge, *Essai pour une histoire des voix au dix-huitième siècle*.

31. Kreilkamp, *Voice and the Victorian Storyteller*, 2.

32. Steven Connor, *Dumbstruck: A Cultural History of Ventriloquism*, Oxford: Oxford University Press, 2000, 43.

33. Tom Delph Janiuriek has argued for an inclusion of the voice in the analyses of performance of gender along the lines of Judith Butler's concept of gender, but this has hardly been taken up outside the confines of the artistic/staged voice. Tom Delph Janiurek, "Sounding Gender(ed): Vocal Performances in English University Teaching Spaces," *Gender, Place and Culture: A Journal of Feminist Geography*, 6:2, (1999), 137–153.

34. Martha Feldman, *The Castrato: Reflections on Natures and Kinds*, Oakland: University of California Press, 2015.

35. Naomi André, *Voicing Gender: Castrati, Travesti, and the Second Woman in Early-Nineteenth-Century Italian Opera*, Bloomington: Indiana University Press, 2006. On the disappearance of the castrato and the rise of the tenor, see also Patrick Barbier, *The World of the Castrati: The History of an Extraordinary Operatic Phenomenon*, London: Souvenir Press, 1998, and John Potter, *Tenor: History of a Voice*, New Haven, CT: Yale University Press, 2009.

36. On the sensuous pleasure of hearing angels cry on the operatic stage, see Michel Poizat, *L'Opera, ou, Le cri de l'ange: Essai sur la jouissance de l'amateur d'opera*, Paris: Métailié, 1986. On the double nature of the diva's body and voice, see James Q. Davies, *Romantic Anatomies of Performance*, 66–92.

37. Christina Luckyj, *"A Moving Rhetoricke": Gender and Silence in Early Modern England*, Manchester: Manchester University Press, 2002, 5.

38. Bloom, *Voice in Motion*, 39.

39. Ann-Christine Mecke, *Mutantenstadl. Der Stimmwechsel und die deutsche Chorpraxis im 18. und 19. Jahrhundert,* Berlin: WVB, 2007.

40. Mecke, *Mutantenstadl,* 249.

41. Philip Brett, Elizabeth Wood, and Gary Thomas, *Queering the Pitch: The New Gay and Lesbian Musicology,* New York: Routledge, 1994.

42. See, e.g., Stan Hawkins (ed.), *Queerness in Pop Music. Aesthetics, Gender Norms, and Temporality,* London: Routledge, 2016.

43. Freya Jarman-Ivens, *Queer Voices: Technologies, Vocalities and the Musical Flaw,* London: Palgrave-Macmillan, 2011.

44. "The pleasure of the acoustic sphere, symbolized by female creatures, sooner or later evokes the figure of the mother," Adriana Cavarero, *For More Than One Voice: Toward a Philosophy of Vocal Expression,* Stanford: Stanford University Press, 2005, 131.

45. Felicia Miller-Frank, *The Mechanical Song: Women, Voice, and the Artificial in Nineteenth-Century French Narrative,* Redwood: Stanford University Press, 1995.

46. Mladen Dolar, *A Voice and Nothing More,* 2006.

47. Rockey, *Speech Disorder,* 14.

48. Joshua St. Pierre, *Cheap Talk: Disability and the Politics of Communication,* Ann Arbor: University of Michigan Press, 2022.

49. Riley McGuire, "The Victorian Unspeakable: Stammering and Same-Sex Intimacy between Men." *DiGeSt. Journal of Diversity and Gender Studies,* 3:2 (2016): 43–57; Daniel Martin, "George Catlin's Shut Your Mouth, the Biopolitics of Voice, and the Problem of the 'Stuttering Indian,'" in Josephine Hoegaerts and Janice Schroeder (eds.), *Ordinary Oralities: Everyday Voices in History,* Oldenbourg: De Gruyter, 2023, 65–80.

50. Jennifer Stoever, *The Sonic Color Line: Race and the Cultural Politics of Listening,* New York: NYU Press, 2016.

51. W. E. B. Du Bois, *The Souls of Black Folk,* New York: Penguin, 1903; Toni Morrison, *Song of Solomon,* New York: Alfred Knopf, 1977.

52. Nina Sun Eidsheim, *The Race of Sound. Listening, Timbre and Vocality in African American Music,* Durham, NC: Duke University Press, 2018, 4.

53. As Dylan Robinson has pointed out, Indigenous methods present alternative modes of knowing and listening, centering various forms of experience rather than modern Western notions of expertise. Robinson, *Hungry Listening: Resonant Theory for Indigenous Sound Studies,* Minneapolis: University of Minnesota Press, 2020.

54. James Stark, *Bel Canto: A History of Vocal Pedagogy,* Toronto: University of Toronto Press, 57–90.

55. Christopher Reid, *Imprison'd Wranglers: The Rhetorical Culture of the House of Commons 1760–1800,* Oxford: Oxford University Press, 2012; Ben Griffin, *The Politics of Gender in Victorian Britain: Masculinity, Political Culture, and the Struggle for Women's Rights,* Cambridge, UK: Cambridge University Press, 2012; Joseph Meisel, *Public Speech and the Culture of Public Life in the Age of Gladstone,* New York: Columbia University Press, 2001.

56. Tom Wright, *Lecturing the Atlantic: Speech, Print, and an Anglo-American Commons 1830–1870,* Oxford: Oxford University Press, 2017; Henk te Velde, *Sprekende politiek. Redenaars en hun publiek in de gouden eeuw,* Amsterdam: Prometheus, 2015; Theo Jung, "Le silence du peuple: The Rhetoric of Silence During the French Revolution," *French History,* 31:4 (2017): 440–469.

57. Ingrid Sykes, "Sounding the Citizen Patient: The Politics of Voice in Post-Revolutionary France," *Medical History,* 55 (2011), 479–502.

58. Josephine Hoegaerts and Janice Schroeder, *Ordinary Oralities: Everyday Voices in History,* Oldenbourg: De Gruyter, 2023, 7.

59. Although there is, of course, meaning to vocal utterances beyond language. See, e.g., Steven Connor, *Beyond Words: Sobs, Hums, Stutters and Other Vocalizations,* Chicago: University of Chicago Press, 2014.

60. The connection between voice and political power is not limited to the context of democracy. In early modern Europe the monarch's voice carried particular meaning (and indeed privilege). See, e.g., Gina Bloom's study of Queen Elizabeth's voice (Bloom, *Voice in Motion,* 187–197) and Bruce Smith's analysis of the soundscape of the English early modern court (Bruce R. Smith, *The Acoustic World of Early Modern England: Attending to the O-factor,* Chicago: University of Chicago Press, 1999, 49–95).

61. Katherine Bergeron, *Voice Lessons: French Mélodie in the Belle Epoque,* Oxford: Oxford University Press, 2010, 83.

62. Studies of the American elocutionary movement are a notable exception. See, e.g., Marian Wilson Kimber, *The Elocutionists: Women, Music, and the Spoken Word*, Urbana: University of Illinois Press, 2017.

63. See, e.g., Liesl Yamaguchi, "Sensuous Linguistics: On Saussure's Synaesthesia," *New Literary History*, 50:1 (2019), 23–42; Tanvi Solanki, "Listening to Difference: J. G. Herder's Aural Theory of Cultural Diversity in the 'Treatise on the Origin of Language' (1772)," *History of European Ideas,* 48:7 (2022), 930–947.

64. In sound studies, listening and the associated audile techniques tend to be stressed over human/physical sound production. See, e.g., Trevor Pinch and Karin Bijsterveld (eds.), *The Oxford Handbook of Sound Studies,* Oxford: Oxford University Press, 2012.

65. Trevor Pinch and Karin Bijsterveld, Introduction to "Sound Studies: New Technologies and Music," Special Issue, *Social Studies of Science,* 34:5 (2004), 636.

66. E.g., Ana Maria Ochoa Gautier, *Aurality: Listening and Knowledge in Nineteenth-Century Colombia*, Durham, NC: Duke University Press, 2014; Joeri Bruyninckx, *Listening in the Field: Recording and the Science of Birdsong*, Cambridge, MA: The MIT Press, 2018.

67. Jonathan Sterne, *The Audible Past,* 7.

68. Sterne, *The Audible Past,* 8.

69. Sterne, *The Audible Past,* 9.

70. Shane Butler, *The Ancient Phonograph,* New York: Zone Books, 2015, 16.

71. Jonathan Rée, *I See a Voice: Deafness, Language and the Senses,* London: Metropolitan Books, 1999, 4.

72. John Picker, *Victorian Soundscapes,* Oxford: Oxford University Press, 2003, 132–135.

73. Nicholas Cook, *Beyond the Score: Music as Performance,* Oxford: Oxford University Press, 2013, 3.

74. Carolyn Abbate, *Unsung Voices: Opera and Musical Narrative in the Nineteenth Century, Princeton, NJ: Princeton University Press, 1991,* 11.

75. Cook, *Beyond the Score,* 6.

76. Shane Butler, *The Ancient Phonograph,* 96.

77. As Karin Martensen has pointed out, this means that recordings effectively co-create vocal practice. Martensen, *"The Phonograph Is Not an Opera House": Quellen und Analysen zu Ästhetik und Geschichte der frühen Tonaufnahme am Beispiel von Edison und Victor,* Munich: Allitera Verlag, 2019.

78. In order to reflect the corpus as a whole, the bibliography at the end of this book contains a list of all primary sources that were consulted and analyzed, rather than only those cited.

Although the list of works does not allow for a quantitative analysis, it does provide some insight into modes of cultural exchange, translation, and the mobility of knowledge. In my overview, some works are mentioned more than once, as they were present in different collections. Where possible, I have retained different versions of works (e.g., in translation). Books with several runs have only been included with the earliest edition present in the collection, unless further editions were substantially revised, in which case they are included as well.

79. Martha Feldman and Judith T. Zeitlin, *The Voice as Something More: Essays Toward Materiality*, Chicago: University of Chicago Press, 2019, xii.

80. Feldman and Zeitlin, *The Voice as Something More*, 9.

81. Roland Barthes, "The Grain of the Voice," in Jonathan Sterne (ed.), *The Sound Studies Reader*, London: Routledge, 2012, 506.

CHAPTER 1

1. Marc Colombat de l'Isère, *Tableau synoptique et statistique de toutes les espèces de bégaiement: et des moyens curatifs qui conviennent à chaque variété, chez les principaux libraires de médecine*, 1833.

2. Marc Colombat de l'Isère, *Tableau synoptique*. "Médecin fondateur et directeur de l'institut orthophonique pour le traitement des bègues et des muets; collaborateur de plusieurs journaux de médecine et scientifiques; membre de la société anatomique de Paris, du comité consultatif de la société pour l' émancipation intellectuelle, du cercle chirurgical de Montpellier, de la société médico-chirurgicale de Lyon, correspondant de plusieurs sociétés philantrophiques et littéraires, etc."

3. James Hunt identified himself as Ph.D., F.S.A., F.R.S.L., F.E.S., for example. James Hunt, *A Treatise on the Cure of Stammering, with a General Account of the Various Systems for the Cure of Impediments in Speech and a Notice of the Life of the Late Thomas Hunt*, Longman, Green, Longman, and Roberts, 1857.

4. On the role of sociability in early-nineteenth century science and medicine, see, e.g., Michael Brown, *Performing Medicine: Medical Culture and Identity in Provincial England, c. 1760–1850*, Manchester: Manchester University Press, 2011.

5. Thomas Gieryn, *Cultural Boundaries of Science*, 27.

6. James Hunt, *Stammering and Stuttering, Their Nature and Treatment*, London: Longman, Green, Longman, and Roberts, 1861.

7. Fortuné Malebouche, *Institution de M. Malebouche: Nouvelle méthode pour guérir le bégaiement*, impr. de Ducessois, 1828.

8. On the place of unconventional and not formally recognized cures in the medical "market," see Frank Huisman, "Shaping the Medical Market: On the Construction of Quackery and Folk Medicine in Dutch Historiography," *Medical History*, 43:3 (1999), 359–375.

9. In some ways, even though the field is now firmly established as a discipline, speech therapy (or speech language pathology) is still only tenuously connected to the medical field, and is taught in or outside medical faculties depending on country or degree program.

10. For example: Alfred Ephraim, *Die Hygiene des Gesanges. Für Sänger und Gesanglehrer dargestellt*, Leipzig: Breitkopf & Härtel, 1899; Achille Gouguenheim and Marcel Lermoyez, *Physiologie de la voix et du chant: hygiène du chanteur, Paris:* A.Delahaye et E. Lecroisnier, 1885; Aimé Riant, *Hygiène de l'orateur*, Paris: J. B. Baillière et fils, 1886; André Castex, *Hygiène de la voix parlée et chantée*, Paris: Masson, 1894; Morell Mackenzie, *The Hygiene of the Vocal Organs: A Practical Handbook for Singers and Speakers*, London: Macmillan, 1886.

11. Marc Colombat de l'Isère, *Traité des maladies et de l'hygiène des organes de la voix*, Manut, 1838.

12. Joseph Collins, for example, identified himself as a "professor of disease of the mind" and "neurologist." Like most of his colleagues, he focused on the inability to speak, rather than speech itself. J. Collins, *The Genesis and Dissolution of the Faculty of Speech: A Clinical and Psychological Study of Aphasia*, New York: Macmillan, 1898.

13. E.g., Benedict Widmann, *Gehör- und Stimmbildung. Eine auf physiologische, psychologische und pädagogische Untersuchungen und Beobachtungen gegründete Anleitung zur Pflege des Gehörs und der Stimme*, Leipzig: Carl Merzeburger, 1899; David C. Taylor, *The Psychology of Singing: A Rational Method of Voice Culture Based on a Scientific Analysis of All Systems, Ancient and Modern*, New York: Macmillan and Co., 1908; Alfred Appelt, *The Real Cause of Stammering and Its Permanent Cure: A Treatise on Psycho-Analytical Lines*, London: Methuen & Co., 1911; Oscar Aronsohn, *Der psychologische Ursprung des Stotterns*, Carl Marhold Verlagsbuchhandlung, 1914; W. Henri Zay, *Practical Psychology of Voice and of Life*, New York: G. Schirmer, 1917.

14. On the laryngoscope and its invention in the 1850s, see Intermezzo 1.

15. Karl Friedrich Liskovius, *Physiologie der menschlichen Stimme für Aerzte und Nichtärzte*, Leipzig: Verlag von Joh. Amb. Barth., 1846; Henri Monro, *On Stammering and Its Treatment*, London: John Churchill, 1850, i. "Introductory Remarks Important to My Non-Medical Reader."

16. R. Norris Wolfenden, *Studies in Pathological Anatomy, Especially in Relation to Laryngeal Neoplasms*, 1888.

17. V. M. Fourcade, *Actualité, par M*** récitée dans ses séances par M. V.-M. Fourcade, artiste dramatique et lyrique, maître en l'art de parler, enseigneur de la parole aux sourds-muets, professeur de prononciation, correcteur de tout vice d'articulation et de toute défectuosité vocale*, imp. de A. Loger, C.-J. Boulay et Cie, 1861.

18. Theodor Jacobs, *Dauernde Heilung für Stotternde*, Basel: Buchdruckerei zum Basler Berichthaus, 1903.

19. J. P. Sandlands, *The Voice and Public Speaking: A Book for All Who Read and Speak in Public*, London: Hodder and Stoughton, 1880; J. Farrar, *The Human Voice and Connected Parts: A Practical Book for Orator, Clergymen, Vocalists and Others*, London: Marshall Japp & Company, 1881; Theodore Schmauk, *The Voice in Speech and Song;* Leo Kofler, *The Art of Breathing as the Bases of Tone-Production: A Book Indispensable to Singers, Elocutionists, Educators, Barristers, Clergymen, and to All Others Desirous of Having a Pleasant Voice and Good Health*, London: J. Curwen & Sons, 1901.

20. Much like Samuel Smiles's *Self Help* these manuals celebrated the individual agency of the middle class man to rise to a position of leadership (and monetary success). Gotthold Carlberg, *Die Kunst, Sänger zu werden. Eine auf Erfahrung begründete Anleitung, besonders für Dilettanten, sich in der Kunst des Gesanges, in Bezug auf Stimme und Vortrag durch Selbstunterricht zu vervollkommnen*, A. Hartleben's Verlag, 1870; L. Klindworth, *Stuttering and How to Cure It: A Practical and Systematical Handbook for Self-Instruction; Containing an Approved Method of Respiration and the Gymnastic of Voice and Speech*, F.Bauermeister, 1896.

21. Richard Cull, *Stammering, Considered with Reference to Its Cure by the Application of Those Laws Which Regulate Utterance: In a Letter Addressed to George Birkbeck*, London: Henry Renshaw, 1835, 14.

22. "On Stammering," *Medico-Chirurgical Review*, Oct. 1836, 319–322.

23. The idea that the field of speech impediments was particularly vulnerable to quackery was shared by Claude Vernet, Stephen de la Madelaine, and F. Angermann, among others. Claude Vernet, *Leçons théoriques et pratiques sur la cause, le siège, la nature, le mécanisme et le traitement du*

bégaiement, ou Méthode naturelle de traitement du bégaiement en 20 leçons, mise à la portée de tout le monde par Mlle Coralie Vernet et Claude Vernet son père, Clermont-Ferrand, 1841, 14; Stephen de la Madelaine, *Théories complètes du chant,* Paris: Amyot, 1852, 70; F. Angermann, *Das Stottern, Sein Wesen und seine Heilung für Ärzte und Pädagogen,* Berlin: Druck und Verlag von Trowitzsch und Sohn, 1853, i.

24. As an elocutionist, Cull's insistence on experience and practice was, in the first half of the nineteenth century, indeed treated with suspicion, and possibly seen as "mere observation" rather than science. Anna Wierzbicka, *Experience, Evidence and Sense: The Hidden Cultural Legacy of English,* Oxford: Oxford University Press, 2010, 8. "On Stammering," *Medico-Chirurgical Review,* Oct. 1836, 322.

25. "Bis dahin aber im Allgemeinen der grössten Charlatanerie Preis gegeben war," Angermann, *Das Stottern,* iii.

26. In fact, Mlle. Vernet appears in the book as a foil upon which Vernet père can reflect his wisdom. The book is written as a dialogue between a young girl asking rather naïve questions about stammering, and her father, whose answers display his great knowledge on the subject. "le charlatanisme, jusqu'à ce jour, a entouré le traitement de cette infirmité." Vernet, *Leçons théorétiques et pratiques,* iv.

27. Their own status as experts was, of course, inherently fragile and fluid as well: the modern notion of expertise "always remained a socially constructed, inherently unstable form of authority." Joris Vandendriesche, Evert Peeters, and Kaat Wils, "Performing Expertise," in Vandendriessche, Peeters, and Wils (eds.), *Scientists' Expertise as Performance,* London: Routledge, 2015, 2.

28. Joseph Poett, *A Practical Treatise on Stammering and Nervous Affections of Speech, Illustrated with Several Cases of Cure, Authenticated by Persons of Distinction & Medical Authorities,* London: Calkin & Budd, 1833, 80.

29. A mixture of emotions that seems to fit quite well into the affective landscape of nineteenth century science in a broader sense, in which disinterested observation and notions of "sympathy" coexisted in the identity of the gentleman scientist. Rob Boddice, *The Science of Sympathy: Morality, Evolution, and Victorian Civilization,* Champaign: University of Illinois Press, 2016.

30. A. M. Hartley, *Observations on the Different Kinds of Impediments of Speech, Particularly Spasmodic Stammering: With Instructions How to Remove Them,* London: Simpkin, Marshall & Co., 1833, 30.

31. Frank Huisman, "Shaping the Medical Market: On the Construction of Quackery and Folk Medicine in Dutch Historiography," 359–375.

32. Huisman, "Shaping the Medical Market," 360.

33. Huisman, "Shaping the Medical Market," 361.

34. Joris Vandendriessche, "Setting Scientific Standards: Publishing in Medical Societies in Nineteenth-Century Belgium," *Bulletin of the History of Medicine,* 88:4 (2014), 626–653.

35. On the role of the construction and performance of a 'persona' in the practice of science in the nineteenth century, see, e.g., Kaat Wils, Mineke Bosch, and Kirstin Niskanen (eds.), Special Issue: Scientific Persona, *Persona Studies,* 4:1 (2018).

36. Fortuné Malebouche, *Institution de M. Malebouche. Nouvelle méthode pour guérir le bégaiement,* 4.

37. Chimène Bateman, "Irrepressible Malebouche: Voice, Citation and Polyphony in the Roman de la rose," *Cahiers de recherches médiévales et humanistes,* 22 (2011), 9–22.

38. Ibid., 15.

39. "Madame Leigh, habitant New-York, devenue veuve à l'âge de 36 ans, fut accueillie avec bienveillance dans la famille du docteur Yates et y reçut les soins les plus désintéressés. Une des filles

de ce médecin, âgée d'environ 18 ans, était atteinte d'un bégaiement assez fort. Madame Leigh crut ne pouvoir mieux témoigner sa reconnaissance à ses hôtes, qu'en délivrant cette demoiselle de son infirmité; elle lut à cet effet tous les ouvrages anglais qui ont trait au bégaiement. Mais n'obtenant pas de cette étude ce qu'elle en désirait, elle se borna à observer avec persévérance la nature de l'infirmité qu'elle voulait guérir, sur le sujet même qui en était affecté," Malebouche, *Institution,* 1828, 14.

40. "L'inventrice de cette méthode, encouragé par le succès qu'elle obtint en Amérique, voulut la répandre en Europe, et en confia le soin à M. Malebouche, frère de celui qui est maintenant à Paris." Ibid., 16.

41. See for example an anonymous song of a "quack," published in the eighteenth century: "You that would last long, list to my song, / make no more coil, but buy of this oil, / would you be ever fair and young, / stout of teeth, and strong of tongue." s.n. The harangues, or speeches, of several celebrated quack-doctors in town and country, 1762.

42. "Vos commissaires déclarent que, par la méthode de la veuve Leigh, on peut parvenir à guérir le bégaiement [...]; mais ils ne peuvent s'empêcher d'exprimer le regret que l'inventeur de cette méthode n'ait pas cru devoir se conformer à l'usage consacré de nos jours, de rendre publiques toutes les découvertes qui peuvent devenir profitables à l'humanité," Malebouche, *Institution,* 1828, 27–28.

43. Ibid., 10.

44. Fortuné Malebouche, *Précis sur les causes du bégaiement et sur les moyens de le guérir,* Paris: Fortin, Masson etc., 1841, 6.

45. Ibid., 10.

46. Ibid.

47. On Dieffenbach and other surgeons searching for a cure for stuttering in the nineteenth century, see Yvan Lebrun and Madeleine Bayle, "Surgery in the Treatment of Stuttering," in Yvan Lebrun and Richard Hoops (eds.), *Neurolinguistic Approaches to Stuttering,* Berlin: Mouton, 1973, 82–90.

48. Malebouche, *Précis,* 1841, 19.

49. Ibid.

50. Malebouche claimed that he sold his methods at state level, e.g., to the government of the Netherlands.

51. "il y a une tendance énorme à ne tenir aucun compte des experiences faites, des précédents avérés, et une disposition non moins marquee à mépriser la théorie," Malebouche, *Précis,* 1841, 14.

52. Ibid., 31.

53. Ibid., 8.

54. Cull, *Stammering,* 14.

55. Ibid.

56. Eugène Paignon, *Eloquence et improvisation: art de la parole oratoire au barreau, à la tribune, à la chaire,* Paris: Cotillon, 1846, 22."Tant que la grande voix de Demosthène put se faire entendre, Athènes fut sauvée. Cicéron parle, Catilina n'est plus. Rome périt quand l'éloquence dégénéra."

57. James Mackness, *Dysphonia Clericorum or Clergyman's Sore-Throat: Its Pathology, Treatment, and Prevention,* London: Longman, Brown, Green, and Longmans, 1848, 2.

58. Ibid.

59. Joseph Meisel, *Public Speech and the Culture of Public Life in the Age of Gladstone,* 11–50.

60. François Bennatti, *Recherches sur le mécanisme de la voix humaine,* J.-B. Baillière, 1832, 112. "Les analyses imparfaites et les traités vagues que les anciens nous ont laissés sur la voix, et plus encore la confusion qui existe dans les travaux de leurs commentateurs, ne nous permettent pas de

décider si les termes os rotundum dont se servaient les Romains pour designer la supériorité de la prononciation des Grecs, se rapportaient à la construction de leurs périodes, à l'abondance ou à la position desvoyelles ou àla qualité de la voix."

61. As Simon Goldhill points out, Plutarch's reputation as a historian and a scholar suffered considerably in the nineteenth century, but he was seen as a typical schoolboy's read—and that is probably how these gentlemen scientists (who were not classicists, after all) learned about Demosthenes. Simon Goldhill, "The Value of Greek: Why Save Plutarch," in *Who Needs Greek? Contests in the Cultural History of Hellenism,* Cambridge, UK: Cambridge University Press, 2002.

62. Charles Rivail, *De l'art de la parole au barreau, disocurs prononcé à la séance solennelle des conférences*, impr. de F. Allier père et fils, 1890, 16. "Dans les lycées, on donne une leçon, on lit un devoir, et c'est tout. Il n'y a pas d'exercice d'improvisation. C'est un tort, car l'art de la parole s'apprend, et, pour être orateur, il faut toujours, plus ou moins, avoir mâché les cailloux de Démosthène."

63. Jean-Baptiste Grognier-Quélus, *Etudes sur l'art dramatique et oratoire: conseils aux comédiens & comédiens chanteurs*, impr. de Detrie-Tomson, 1858, 62. "Ce prince des orateurs, malgré la puissance de son génie et la vigueur de son élocution, ne fut bien accueilli que lorsqu'il eut compris que, sans l'action, les plus belles choses ne sont qu'un corps sans vie, plus propres à glacer l'auditeur qu'à l'échauffer."

64. Louis Dubroca, *Principes raisonnés sur l'art de lire à haute voix, suivis de leur application particulière à la lecture des ouvrages d'éloquence et de poésie,* Paris: Imprimerie de Farge, 1802, 301. "La réponse de Démosthènes, citée par Cicéron et par Quintilien, nous apprend l'importance que cet éloquent orateur attachait à la prononciation et à son influence. Interrogé sur la question de savoir quel était le ressort principal de l'art oratoire C'est l'action, répondit-il; et le second, ajouta-t-on?: C'est l'action et le troisième? C'est l'action, répondit toujours le prince des orateurs Grecs. Il n'est pas étonnant qu'il ait apprécié à ce point cette partie de l'éloquence, et qu'il ait fait tant d'efforts pour y atteindre; car elle est incontestablement de la plus haute importance."

65. Henry McCormac, *A Treatise on the Cause and Cure of Hesitation of Speech, or Stammering,* London: Longman, Rees, Orme, Brown, and Green, 1828, 75.

66. Mansuès-François Ramont, *De la voix et de la parole,* Paris: Imprimerie de Feugueray, 1803, 125. "Demosthènes apporta en naissant cette fâcheuse disposition, mais il parvint dans la suite à rectifier et perfectionnersa prononciation, en s'arrêtant tous les jours sur le rivage de la mer, où, la bouche remplie de cailloux, il haranguait les flots."

67. Félix Voisin, *Du Bégaiement, ses causes, ses différens degrés, influence des passions, des sexes, des âges, etc. sur ce vice de prononciation; moyens thérapeutiques pour prévenir, modifier ou guérir cette infirmité,* Paris: Croullebois, l'auteur, 1821, 39. "L'exemple tant de fois cité de Démosthène doit soutenir leur courage et les exciter à tenter tous les moyens pour donner aux mouvemens des organes de la parole la force et la précision qui leur manquent. Plutarque raconte avec détail et de la manière la plus piquante comment ce grand orateur parvint à se délivrer d'un bégaiement qui semblait devoir l'exclure pour toujours de la tribune aux harangues. Je crois utile de consigner ici ce passage intéressant. Il doit servir à tous ceux que la nature ou une éducation vicieuse ont frappés du même vice de prononciation."

68. Eduoard Mennechet, *Etudes sur la lecture à haute voix,* Paris: Langlois et Leclerq, 1855, 116. "L'exemple de Démosthène, triomphant, par une volonté ferme et persévérante, de la faiblesse de sa voix et des vices de sa prononciation, ne saurait être mis trop souvent sous les yeux des personnes qui ont le désir d'apprendre àbien lire et à bien parler."

69. Claude Vernet and Coralie Vernet, *Leçons théoriques et pratiques sur la cause, le siège, la nature, le mécanisme et le traitement du bégaiement, ou Méthode naturelle de traitement du bégaiement en 20 leçons, mise à la portée de tout le monde par Mlle Coralie Vernet et Claude Vernet son père,*

16. "CORALIE. Papa, est-il vrai que Démosthène parvint à se guérir par un moyen si simple? Ce n'est guère probable! R. Mon enfant, tout le monde a cru à ce moyen de guérison, et on y croira toujours, si la méthode naturelle n'est pas comprise; mais quiconque saura saisir la cause organique et le mécanisme du bégaiement, reconnaîtra aisément que les petits cailloux dont Démosthène fit usage ne furent que des moyens secondaires ou éphémères dans le traitement, et que l'habitude de parler seul dans la solitude et de s'y exercer d'une manière ardente et soutenue, jointe au régime sévère que Démosthène fut forcé d'observer en vivant dans les bois et sur les rocher escarpés, firent tous les frais de cette belle cure. Si vous doutiez de ce que j'ose avancer, mon enfant, réfléchissez que la plupart des bègues parlent très-bien quand ils sont seuls, et je puis vous assurer que je ne bégaie jamais quand je suis seul, soit que je parle, soit que je lise; je fais ce que je veux de ma langue, et si vous venez à paraître, je suis troublé: cependant votre présence m'est bien agréable; il est vrai que la difficulté que j'éprouve est très-légère; mais encore elle a lieu. Chose étrange! que la présence d'un autre soi-même fasse naître le bégaiement! Aussi Démosthène sut-il profiter de cette circonstance, et se hâta-t-il de prendre le sage parti de fuir la société, et de se retirer dans la solitude; là, livré à lui-même, il s'exerça long-temps à l'usage de la parole et finit par perdre la vicieuse habitude de bégayer!"

70. Quintilian, *Institutione Oratoria*, Chap.1.

71. J. Duquesnois, *Manuel de l'orateur et du lecteur*, Paris: Chez Vaton, 1841, 111. "Il est peut-être encore plus urgent d'introduire dans les établissements de demoiselles les fondements d'une bonne prononciation et d'une bonne lecture. En effet, les demoiselles ne sont-elles pas appelées à devenir mères, et comme telles à diriger l'enfance? 'C'est du soin qu'elles prennent à former nos premiers sons que depend pour nous une bonne ou une mauvaise prononciation', dit Quintilien."

72. Auguste Debay, *Hygiène de la voix et gymnastique des organes vocaux*, Paris: Moquet, 1852, 172. "L'art vocal était devenu à Rome une profession aussi lucrative qu'honorée."

73. James Mackness, *Dysphonia Clericorum 'or Clergyman's Sore-Throat*, 120.

74. Édouard Fournié, *Physiologie de la voix et de la parole. Avec figures dans le texte*, Paris: Adrien Delahaye, Libraire-éditeur, 1866, 204. "Cela est très-faux, mais c'est éminemment logique."

75. T. Bartlett, *Stammering Practically Considered with the Treatment in Detail*, London: Sherwood, Gilbert and Piper, 1839, 19–20.

76. On nineteenth-century research on acoustics, see, e.g., Thomas Rossing, "A Brief History of Acoustics" in Rossing (ed.), *Handbook of Acoustics*, New York: Springer, 2007, 9–24.

77. Fournié, *Physiologie de la voix*, 201. "il n'a pas été question de la formation des tons, ni des modifications de la voix, selon les sexes, l'âge, l'état de santé ou de maladie. Cette lacune se trouve comblée d'une manière assez satisfaisante dans les livres que nous allons examiner."

78. Fournié, *Physiologie de la voix*, 177. "Historique et critique des différentes théories de la voix."

79. Fournié, 177. "L'extension peu commune que nous avons donnée à la partie historique et critique de notre travail mérite d'être justifiée."

80. Fournié, 178, "La tâche d'historien critique n'est pas toujours facile."

81. Fournié, 178, "L'on ne saurait, en effet, rassembler trop de documents historiques sur toutes les questions dans un moment où, pour répondre à un besoin general, l'on est occupé à édifier une histoire de la médecine sur des bases plus intelligentes et plus rationnelles."

82. He noted differences according to sex and age, for example.

83. Fournié, 219. "Après Galien, il faut arriver jusau'à Fabrice d'Aquapendente pour trouver un auteur que se soit sérieusement occupé des organes de la voix."

84. Fournié, 223

85. Fournié, 246. "Mersenne s'était pénétré profondément des travaux des anciens."

86. Fournié, 295. "Les savants de notre époque."

87. Fournié, 349. "Nos investigations se sont étendues, comme on l'a vu jusqu'ici, aussi bien sur la partie physique que sur la partie anatomique; le maniement du laryngoscope nous est assez familier; nous avons cherché, en un mot, à réunir toutes les conditions favorable à la production d'un travail consciencieux et utile."

88. Wolfgang von Kempelen, *Mechanismus der menschlichen Sprache: nebst der Beschreibung seiner sprechenden Maschine*, J. V. Degen, 1791, 399.

89. William Gardiner, *The Music of Nature or an Attempt to Prove That What Is Passionate and Pleasing in the Art of Singing, Speaking, and Performing upon Musical Instruments Is Derived from the Sounds of the Animated World. With Curious and Interesting Illustrations,* London: Longman, Rees, Orme, Brown, Green and Longman, 1832.

90. Roland Barthes, "The Grain of the Voice," Jonathan Sterne (ed.), *The Sound Studies Reader,* 2012.

91. On reporting practices in parliament, see, e.g., Andrew Sparrow, *Obscure Scribblers: A History of Parliamentary Journalism*, London: Politico, 2003.

92. Josephine Hoegaerts, "The Sound of Sovereignty: Royal Vocal Strategies in the Victorian House of Lords" in Anke Gilleir and Aude Defurne (eds.), *Strategic Imaginations: Women and the Gender of Sovereignty in European Culture,* Leuven: Leuven University Press, 2020.

93. This is demonstrated most clearly, perhaps, by Bruce Smith as he exhorts readers to "attend to the O-factor" in his history of the sounding reality of early modern practices of politics and theatre. Bruce Smith, *The Acoustic World of Early-Modern England,* 3–29.

94. Mark Delaere, "De toekomst van de muziekwetenschap: wat is een partituur?," *Tijdschrift voor Muziektheorie,* 11:1 (2006), 31–35.

95. Meanwhile, speakers in parliament were increasingly anxious about what they perceived as too much accuracy in reporting (for example, with the use of shorthand, which allowed for an almost completely literal transcription of the spoken word). Tension between speakers' desire to be represented as fluent and grammatical and a heightened sense of technically accurate reproduction of sounds would continued throughout the century. See, e.g., Josephine Hoegaerts, "Speaking Like Intelligent Men: Vocal Articulations of Authority and Identity in the House of Commons in the Nineteenth Century," *Radical History Review* (themed issue: Sound Politics), 121 (2015), 123–144.

96. Charles Bataille, *Nouvelles recherches sur la phonation: Mémoire présenté et lu à l'académie des sciences le 15 avril*, Paris: Victor Masson et fils, 1861.

INTERMEZZO I

Epigraph: George Johnson, *The Laryngoscope. Directions for Its Use, and Practical Illustrations of Its Value in the Diagnosis and Treatment of Diseases of the Throat and Nose: Two Lectures Delivered at the Royal College of Physicians,* London: Robert Hardwicke, 1864, 12.

1. Ibid., 13.

2. James Prosser, *Lessons in Laryngoscopy Including Rhinoscopy and the Diagnosis and Treatment of Diseases of the Throat*, London: Baillier, Tindall and Cox, 1873, 84.

3. Horace Green (1802–1866) was an American physician who specialized in diseases of the larynx, pharynx, and trachea and was mainly active in New York. The American Broncho-Esophagological Association credits him with the first performance of direct laryngoscopy. Johann Nepomuk Czermak (1828–1873) was an Austrian-German physiologist who established the Leipzig *Spectatorium,* and specialized in studying phonetics. Together with Viennese neurologist Ludwig Türck (1810–1868), he played an important part in introducing laryngoscopy into mainstream medicine.

4. Archibald E. Garrod, *An Introduction to the Use of the Laryngoscope*, London: Longman, Green & Co., 1886, 1; Johnson, *The Laryngoscope*, 12; Morrell Mackenzie, *The Use of the Laryngoscope in Diseases of the Throat. With an Appendix on Rhinoscopy*, London: Robert Hardwicke, 1865, 25.

5. Malcolm Sterling Mackinlay, *Garcia the Centenarian and His Time; Being a Memoir of Manuel Garcia's Life and Labours for the Advancement of Music and Science*, London: Hutchinson & Co., 1908, 201–213.

6. Ludwig Türck, *Praktische Anleitung zur Laryngoskopie. Mit 32 Holzschniten und einer Steindrucktafel*, Vienna: Wilhelm Braumüller, 1860.

7. George Duncan Gibb, *The Laryngoscope: Illustrations of Its Practical Application, and Description of Its Mechanism*, London: John Churchill, New Burlington Street, 1863, 5.

8. Mackenzie, *The Use of the Laryngoscope*, 25–34.

9. Ibid., vii.

10. Ibid., 1.

11. Ibid., 26.

12. Ibid.

13. Mackenzie, *The Use of the Laryngoscope*, 31.

14. Ibid., 32.

15. Sterling Mackinlay, *Garcia the Centenarian and His Time*, 203.

16. Ibid., 205.

17. Ibid., 206.

18. Ibid., 201.

19. Ibid., 212.

20. Ibid., 204.

21. On the *coup de la glotte* and the controversies surrounding it, see James Stark, *Bel Canto: A History of Vocal Pedagogy*, 12–19.

22. Mathilde Marchesi, *Marchesi and Music: Passages from the Life of a Famous Singing Teacher*, New York: Harper & Brother Publishers, 1897, 25.

CHAPTER 2

1. See, e.g., Gertrude Himmelfarb, "Manners into Morals: What the Victorians Knew," *American Scholar*, 57:2 (1988), 223–232.

2. See, e.g., Jürgen Kocka, "Bürgertum und Bürgerlichkeit als Probleme der deutschen Geschichte vom späten 18. zum frühen 20. Jahrhundert," in J. Kocka (ed.), *Bürger und Bürgerlichkeit im 19. Jahrhundert*, Göttingen: Vandenhoeck & Ruprecht, 1987.

3. The *histoire des sensibilités* covered themes like fear and love, but also smell and touch, and was mostly developed in the 1980s and 1990s by historians like Alain Corbin, Jean Delumeau, Georges Vigarellos, and others.

4. Martin Hewitt, *Making Social Knowledge in the Victorian City*; Ute Frevert, *Gefühle in der Geschichte. Kritische Studien zur Geschichtswissenschaft*, Göttingen: Vandenhoeck & Ruprecht, 2021; Delphine Gardey, *Le linge du Palais Bourbon: Corps, materialité et genre du politique á l'ère démocratique*, Lormont: Éditions Le Bord de l'Eau, 2015.

5. Jennifer Richards, *Voices and Books in the English Renaissance: A New History of Reading*, Oxford University Press, 2019, 18.

6. Joy Damousi, *Colonial Voices: A Cultural History of English in Australia 1840–1940, Cambridge, UK:* Cambridge University Press, 2010, 15–16.

7. The fixation of identity markers like gender and ethnicity into biological universals at the turn of the nineteenth century has been remarked upon in numerous studies, e.g., Thomas Laqueur, *Making Sex: Body and Gender from the Greeks to Freud,* Cambridge, MA: Harvard University Press, 1992; Gail Bederman, *Manliness and Civilization. A Cultural History of Gender and Race in the United States, 1880–1917,* Chicago: University of Chicago Press, 1995.

8. Josephine Hoegaerts, "By but Not of the Body: Performing Girls' and Boys' Voices in the Nineteenth Century," in Heidrun Zettelbauer and Stefan Benedik (eds.), *Embodiment—Verkörperung,* Göttingen: Vandenhoeck & Ruprecht, 2017.

9. Joseph Mainzer, *Méthode de chant pour les enfants à l'usage des écoles et des pensions,* Paris: Chez Maurice Schlesinger, 1836, xiii. "A une époque enfin où ces organes sont encore souples et accessibles à toutes les impressions."

10. *A Treatise on the Nature and Causes of Stammering, with an Exposition of the Best Methods of Cure, Medical, Surgical, and Educational,* by a physician, London: Samuel Highly, 1843, 35.

11. Christopher Rowley, *The Voice: or the Physiologist Versus the Singing Master,* London: Office of Music, 1898.

12. Peter Burke, *The Art of Conversation,* London: Polity, 1993.

13. *The Birmingham Journal,* 5

14. William Gardiner, *The Music of Nature,* 311.

15. See, e.g., L. Grünwald, *Atlas and Abstract of the Diseases of the Larynx,* London: Rebman Publishing Co., 1898, 27; Colombat de l'Isère, *L'orthophonie ou physiologie et thérapeutique du bégaiement et de tous les vices de la prononciation,* Paris: Mansut fils, 1833, 84.

16. Édouard Mennechet, *Etudes sur la lecture à haute voix,* 65.

17. William Sleight, *A Voice from the Dumb: A Memoir of John William Lashford, Late Pupil in the Brighton and Sussex Institution for the Deaf and Dumb,* London: Hamilton, Adams, and Co., 1849.

18. William Sleight, *A Voice from the Dumb,* 14.

19. As numerous deaf studies scholars have shown, the widespread preference in nineteenth-century Europe for oralist education (especially among hearing educators) led to excessive attention to deaf voices, and their assimilation into hearing society—and was rooted in audism. See, e.g., Harlan Lane, *The Mask of Benevolence: Disabling the Deaf Community, San Diego:* Dawn Sign Press, 1992.

20. Hippolyte-Pierre Crosilhes, *Hygiène et maladies de la poitrine et de la voix,* Paris: Moquet, 5.

21. Joseph-Ferdinand Bernard, *Gymnastique pulmonaire, ou l'art de respirer pour chanter, parler et vivre,* Lyon: impr. de Jevain et Bourgeon, 1868, 12. "Toutes ceintures, corsets et vêtements serrés sont nuisibles aux exercices et à la santé en general."

22. Josephine Hoegaerts, "Women's Voices in Educational Manuals: The Gendered Sounds of Speech Therapy, Song and Education in Europe, c. 1830–1900," *Women's History Review* (2019).

23. Notably, the rising popularity of physical culture and bodybuilding was often driven by narratives about slight, weak men transforming themselves into strongmen.

24. Mackness, *Dysphonia clericorum,* 25.

25. Mackness, *Dysphonia clericorum,* 96.

26. Joy Damousi notes a similar association of female voices with a natural affinity for civilized sounds in the settler colonial context of Australia. Damousi, *Colonial Voices.* In the second half of the nineteenth century, this would start to change somewhat as women increasingly took to the stage with the spoken word as well, crossing boundaries between domestic and public settings, and between music and speech. See Marian Wilson Kimber, *The Elocutionists.*

27. A. de B., "Les Jeunes Cantatrices," *Gazette Musicale de Paris,* 40: (2),1835, 321–323, "Dans quel répertoire trouverez-vous, s'il vous plaît, des rôles écrits, je ne dis pas pour des voix usées, cela ne

peut venir à l'idée d'un compositeur raisonnable, mais pour des voix qui ne sont déjà plus dans tout l'éclat de leur sonorité? La passion musicale est tout extérieure, tout ardente, tout exagérée; elle ne discute pas, elle agit. Pour agire, il faut être jeune; la passion musicale ne peut se passer de jeunesse."

28. Ad. Gueroult, "De l'enseignement populaire du chant: Réunion générale des cinq divisions de l>orphéon," *Gazette Musicale de Paris,* 1839, 137–139. "La belle idée, disait-elle, de leur apprendre de la musique! elles ont bien besoin de cela pour faire leur cuisine et pour nettoyer leurs enfants! Ce sera une pépinière d'actrices. *Heureusement* qu'elles sont trop laides pour cela!"

29. Ibid., "Eh! pourquoi donc, madame, tout en remuant vos casseroles, votre cuisinière ne chanterait-elle pas? Eh! où serait le mal, si, au lieu de chanter faux, elle chantait juste? . . . laissez le peuple apprendre la musique, et remerciez, croyez-moi ceux qui lui enseignent à chanter, car rien ne réjouit l'âme comme de chanter."

30. Laurent de Rillé, *Du chant choral,* Paris: Perrotin, editeur de l'orpheon, 1856, 1. "Lorsque la tribu tout entière, lorsque la nation est émue à son tour, la tribu, la nation, s'assemble aussi, et des centaines, des milliers de voix unies, entonnent un hymne religieux ou Guerrier."

31. De Rillé, *Du chant choral,* 12. "Le vrai but des sociétés chorales est de donner à leurs membres des habitudes d'ordre en leur faisant accepter et pratiquer le principe de la solidarité volontaire, et de répandre dans le pays le goût de la musique populaire, les germes de l'instruction musical."

32. De Rillé, *Du chant choral,* 37. "Il faut dans un chœur, comme dans une ville assiégée, supprimer toutes les bouches inutiles."

33. E.g., Albert Lemoine, *De la physionomie et de la parole,* Paris: G. Baillière, 1865, 150; and Louis Dubroca, *Principes raisonnés sur l'art de lire à haute voix, suivis de leur application particulière à la lecture des ouvrages d'éloquence et de poésie,* 189.

34. Rebecca Grotjahn, "Seelenvolle Maschine: Natur und Technik im Gesangsdiskurs," in *Naturalezza | Simplicité: Natürlichkeit im Musiktheater,* ed. Vera Grund, Claire Genewein and Hans Georg Nicklaus, Bielefeld: Transcript, 2019, 81–107.

35. See, e.g., Martha Feldman, *The Castrato.*

36. François Joseph Fétis, *La musique mise à la portée de tout le monde,* Paris: Chez Paulin, 1836, 180. "Si c'est un triomphe pour la morale que l'humanité ne soit plus soumise à ces honteuses mutilations, c'est une calamité our l'art que d'être privé de ces voix admirables."

37. Earl Curzon of Kedleston, *Modern Parliamentary Eloquence,* 1911.

38. William Shakespeare, *The Art of Singing: Based on the Principles of the Old Italian Singing-Masters, and Dealing with Breath-Control, Production of the Voice and Registers, Together with Exercises,* London: Metzler & Co., 1899, v.

39. AJ/37/84, 5a, "J'ai l'honneur de vous rappeler que le conseil supérieur saisi d'une demande analogue par M. le Docteur Glover s'est à l'unanimité prononcée contre toute creation de course scientifique (physiologie de la voix, phonation, hygiene vocale etc) dans le conservatoire national ou l'enseignement doit être et rester purement artistique."

40. Leopoldine Ziska, *Handbook of the Four Elements of Vocalization.* Written and dedicated to Lady Kerrison, London: Ewer &Co., 1850 (edited by Leopold Wray), 17.

41. The so-called Swedish system of gymnastic or calisthenics, particularly, claimed to be based on natural movement, and to promote health. See, e.g., Anders Ottosson, "The First Historical Movements of Kinesiology: Scientification in the Borderline Between Physical Culture and Medicine Around 1850," *International Journal of the History of Sport,* 27:11 (2010), 1892–1919.

42. Félix Voisin, *Du bégaiement, ses causes, ses différens degrés, influence des passions, des sexes, des âges, etc sur ce vice de prononciation,* Paris: Croullebois, l'auteur, 1821.

43. On the development of treble voices, see, e.g., Ann Christine Mecke, *Mutantenstadl: Der Stimmwechsel und die deutsche Chorpraxis im 18. und 19. Jahrhundert.* The nineteenth-century idea

of innocence and the natural quality in the child's voice has its roots in older ideas about treble voices; see, e.g., Hans Georg Nicklaus, "Stimmen der Unschuld. Kastratengesang im Kontext von Kindheit und Pädagogik," in *Naturalezza | Simplicité: Natürlichkeit im Musiktheater*, ed. Vera Grund, Claire Genewein and Hans Georg Nicklaus, Bielefeld: Transcript, 2019.

44. Édouard Fournié, *Physiologie de la voix et de la parole. Avec figures dans le texte,* 540. "Dominée par sa destinée nouvelle, la jeune fille semble déjà prévoir le rôle important que l'utérus va jouer dans sons existence physiologique et pathologique. Triste, rêveuse, le regard langoureux et voilé, elle recherche la solitude pour s'écouter vivre et caresser des sensations intérieures qu'elle ne connaissait pas."

45. Fournié, 538. "Jusqu'à l'âge de treize à quatorze ans, la voix de la jeune fille diffère peu de la voix du jeune garçon. Le timbre et le diapason sont à peu près les mêmes dans les deux sexes. Mais à partir de ce moment, des dissemblances vont surgir. Le jeune garçon va devenir homme dans sa voix, comme la jeune fille deviendra femme dans la sienne. Ce moment suprême est l'âge de la puberté."

46. Carl Ludwig Merkel, *Anatomie und Physiologie des menschlichen Stimm- und Sprachorgans (Anthropophonik),* Leipzig: Verlag von Ambrosius Abel, 1857, 581.

47. T. Bartlett, *Stammering Practically Considered, with the Treatment in Detail,* 22.

48. Naomi André, *Voicing Gender.*

49. John Potter, *Tenor: History of a Voice.*

50. Stephen F. Austin, "'Like the Squawk of a Capon'—the Tenor Do Di Petto," *Journal of Singing* 61:3 (January/February 2005), 309.

51. See, e.g., Jarman, *Queer Voices;* Grotjahn, "Seelenvolle Maschine."

52. Joseph Mainzer, *Méthode de chant pour voix d'hommes à l'usage des collèges, écoles normales, écoles militaires et cours d'ouvriers,* Paris, 1838, vi. "C'est le chant, uni à des paroles à la portée de l'intelligence de la jeunesse, de cet âge de poésie, quie forme le coeur, y fait germer les bons principes, et y grave en caractères ineffaçables les sentimens généreux."

53. Eidsheim, *The Race of Sound.*

54. L.A. Segond, *Mémoire sur l'ossification des cartilages du larynx,* Paris: Rignoux, 1847.

55. Fournié, *Physiologie de la voix et de la parole. Avec figures dans le texte,* 149–151.

56. Colombat de l'Isère, *A Treatise upon the Diseases and Hygiene of the Organs of the Voice,* 13.

57. For more on the mapping of the "ages of man" onto voices, and masculine voices in particular, see Josephine Hoegaerts, "Acquiring a 'Manly' Tone: Audible Intersections of Masculinity and Age in the Long Nineteenth Century," *Genesis: revista della società italiana delle storiche,* 19, 2 (2020).

58. It is interesting to note that the entrance of larger numbers of women into public life in the second half of the twentieth century seems to have led to a drop in pitch in feminine voice, which perhaps denotes the endurance of beliefs around the vocal sound of "maturity" and vocal suitability for professional and public speech. Cecila Pemberton, Paul McCormack, Allison Russell, "Have Women's Voices Lowered Across Time? A Cross Sectional Study of Australian Women's Voices," *Journal of Voice,* 12:2 (1998), 208–213.

59. Colombat de l'Isère, *A Treatise,* 1857, 9.

60. *Hamilton's Modern Instructions in Singing. A Complete Compendium of the Rudiments of Music, Full Directions for the Cultivation and Management of the Voice, Numerous Exercises and Examples on Vocalization, with Solfeggi by Rossini and Sola and a Selection,* London: Robert Cocks & Co., 1853.

61. A. M. Hartley, *Observations on the Different Kinds of Impediments of Speech, Particularly Spasmodic Stammering,* 9.

62. This discourse would become more nuanced later in the nineteenth century, when for example Kate Emil Behnke suggested that the difference between male and female voices was on

average only a fifth, and that alto and tenor voices were, in fact, so closely related that they were often hard to distinguish from each other.

63. Stephen de la Madelaine, *Théories complètes du chant.* "Il suffira de faire remarquer que l'homme chante sur une octave plus basse que la femme."

64. James Mackness, *Dysphonia Clericorum*, 3.

65. Mladen Dolar even imagines the voice as a "bodily missile," launched from the vocalizer's body to, potentially, invade spaces and ears. Dolar, *A Voice and Nothing More,* 70–73.

66. In modern Europe, this was mostly ascribed to a sole Christian God, but ongoing debates about the Apollonian or Dionysian quality of music show a longer and more complicated history of the idea of divine origins of music.

67. Eduoard Mennechet, *Etudes sur la lecture à haute voix,* 34. "Tous les instruments de musique qui charment nos oreilles ont été créés pour ressembler à l'organe de la voix; mais l'organe de la voix ne ressemble qu'à lui-même, c'est l'œuvre de Dieu. L'homme ne peut le comprendre, comment pourrait-il l'imiter?"

68. As Victoria Tkaczyck's extensive work on the history of acoustic science shows, the field was developing significantly in this period. See, e.g., Tkaczyck, "Whose Larynx Is It? Fields of Scholarly Competence Around 1900," *History of Humanities* 3:1 (2018): 57–73, and François Ribac and Viktoria Tkaczyk, "Research on the History of Modern Acoustics," *Revue d'Anthropologie des Connaissances, Société d'Anthropologie des Connaissances,* 2019, *Musical Knowledge, Science Studies, and Resonances,* 13:3, 707–720.

69. Achille Gouguenheim and Marcel Lermoyez, *Physiologie de la voix et du chant: hygiène du chanteur,* 15.

70. Ibid., 19.

71. Marc Colombat de l'Isère, *L'orthophonie ou physiologie et thérapeutique du bégaiement et de tous les vices de la prononciation,* 39.

72. Julien Duquesnois, *Manuel de l'orateur et du lecteur,* 13. "Il me sera facile maintenant, d'après cette définition, de comparer ces différents organes qui constituent la voix à un instrument à vent, tel que la clarinette, qui, lorsque son embouchure est serrée par les dents et les lèvres, produit des sons plus ou moins forts, que les doigts varient en changeant de position. Ainsi la glotte, en ne laissant sortir que peu ou beaucoup de son, en selon qu'il est nécessaire, remplit le même office que les lèvres et les dents placées sur l'anche de l'embouchure de la clarinette, et la langue et les lèvres, en changeant de position, en varient les sons à l'infini, de même que le font les doigts sur le corps de l'instrument."

73. Claudia Schweitzer's history of representations of the human voice as instrument (mostly in France) traces the history of this discourse through the seventeenth and eighteenth centuries, uncovering some epistemological shifts, but no clear chronology in the conceptualization of the larynx. Schweitzer, "La voix humaine: instrument à vent ou instrument à cordes? Un parcours historique (XVIIe/XVIIIe siècles) autour de cette question," *Language & History,* 63:3 (2020).

74. French and German linguists were preoccupied with the connection between language and national identity; see, e.g., Jean Louis Chiss, "Les linguistes du XIXe siècle, l'«identité nationale» et la question de la langue," *Langages*, 182:2 (2011), 41–53.

75. Davies, *Romantic Anatomies of Performance.*

76. Cf. Davies, *Romantic Anatomies*, 66–92, Wayne Koestenbaum, *The Queen's Throat. Opera, Homosexuality and the Mystery of Desire,* New York: Da Capo Press, 1993; Susan Rutherford, *The Prima Donna and Opera, 1815–1930,* Cambridge, UK: Cambridge University Press, 2006.

77. Research on the influence of the operatic "diva" on their audiences and fans bears out the rise of such stars at roughly the same time that ideas about "magnetism" started to take hold in scientific discourse.

78. On the nineteenth century notion of the color of sounds and vowels, see, e.g., Liesl Yamaguchi, "Sensuous Linguistics: On Saussure's Synaesthesia," 23–42.

79. Legouvé is now mostly remembered for his groundbreaking work on women's rights. In the nineteenth century, he was an influential speaker and essayist, and his work—often typified by witty, pithy statements—found its way into numerous publications on the voice in French and English. W. E. Haslam, *Style in Singing*, New York: G. Schirmer, 1911, 15.

80. As Marie Buscatto has shown, female jazz singers in particular had to carve out a place for their voices in a mostly male environment that was largely ruled by ideals of instrumental expertise. Marie Buscatto, *Femmes du jazz,* Paris: Editions CNRS, 2018.

81. Haslam, *Style in Singing,* 16.

INTERMEZZO 2

1. "Dies ist nämlich der allgemeine Glauben unter ihnen: ein Mädchen ginge nur nach Indien, um einen Gatten zu finden. Diese Idee ist so fest angenommen, dass es umsonst ist, zu hoffen, sie würden andere Ursachen annehmen. Und da es auch Brauch unter den Engländern ist, ihre überflüssigen und veralteten Töchter und Schwestern dorthin zu Markte zu senden, so findet es sich meistens, dass diese Frauenzimmer nicht schnell genug mit ihrem Eroberungsplan anfangen können und, sobald sie auf das Schiff kommen, ihre Batterien errichten, um offensiv und defensiv Krieg zu führen." Manfred Elsberger, *Nina d'Aubigny von Engelbrunner. Eine adelige Musikpädagogin am Übergang vom 18. zum 19. Jahrhundert,* BUCH and medi@, 2000, 48.

2. "Nun sollte man glauben, man müsse sehr viel Unterhaltung finden, weil all diese Gegenstände über Asien doch interessant wären. Allein, ich habe noch niemand gefunden, der viel mehr als von den Palaquins und den Tigeranekdoten zu erzählen weiss." Ibid., 51.

3. A similar point was made in the *Gazette Musicale de Paris* in the first half of the nineteenth century, in which journalist and music critic Adolphe Guéroult noted that only "des esprits bicornus" would assume that poor young girls' sole future was in the kitchen and a musical education would be of no use to them. Similar to d'Aubigny, Guéroult imagined young girls as the future adult women who, with their well-developed voices and musical tastes, would be capable of uplifting the cultural well-being of the people in general. In his article, he painted a scene of maids, kitchen aides and nannies singing beautifully and above all cheerfully, rather than gossiping about their employers. "De l'enseignement populaire du chant. Réunion générale des cinq divisions de l'orphéon," *Gazette Musicale de Paris,* 1839.

4. "Aber wenn wir aus dieser Seite unendlich gewonnen haben, so bleibt uns in Ansehung der Bildung der organe, noch ein unerhörter Mangel.," *Briefe an Natalie, 1803,* 3.

5. "Briefe an Natalie über den Gesang als Beförderung der häuslichen Glückseligkeit und des geselligen Vergnügens. Ein Handbuch für Freunde des Gesanges, die sich selbst, oder für Mütter und Erzieherinnen, die ihre Zöglinge für diese Kunst bilden möchten."

6. The epistolary novel (and its nonfictional, autobiographical counterpart) was particularly pervasive in the eighteenth century, and the genre was practiced by women writers as well, which may have inspired d'Aubigny. Elizabeth Heckendorn Cook, *Epistolary Bodies: Gender and Genre in the 18th Century Republic of Letters,* Redwood City: Stanford University Press, 1996.

7. "Dem es wahrlich nicht an Liebe für den Gesang, aber wohl an zweckmässigem Unterricht fehlt idem," 5.

8. Miriam Noa, *Volkstümlichkeit und Nationbuilding. Zum Einfluss der Musik auf den Einigungsprozess der deutschen Nation im 19. Jhdt, New York:* Waxman, 2013, 183–289.

9. *Briefe an Natalie,* 20.

10. "Es ist warlich nur eine Vernachlässigung der schönsten Gabe, wenn man so oft sagen hört: ich habe keine Stimme. Hat nicht die gütige Natur jedem Wesen eine Stimme zum Reden geschenkt? Singt nicht jedes freies Kind auf der Strasse, wenn es frohes Muthes ist? Und wo wäre denn der bestimmte Zeitraum, wo, nach den Gesetzen dieser gütigen Natur, so viele Menschen ihre Stimmen einbüssen müssten, um sie nie wieder zu erhalten, wie man, nach dem seltenen Gebrauch der Stimme beim Gesang, zu urtheilen verleitet werden könnte?

"Wenn man ein Kind seit seinen ersten Jahren ganz von dem Gebrauch seiner Füsse abhalten wollte, so würde es vermuthlich einst eben so richt sagen, ich habe keine Füsse! Wie dies manche Modedame, deren Füssen durch die Füsse ihrer Pferde ersetzt werden, stillschweigend einzugestehen scheint, und die Türken es auf ihren Polstern beweisen, wenn sie den ohne besondern Zweck herum wandelnden Europäer für krank oder für thöricht ansehen." *Briefe an Natalie,* 7.

11. This makes it stand out all the more from other manuals of the time, as most scientific and educational works were written with an implied male reader.

12. "Eingreifend in die physische und moralische Erziehung für das weibliche Geschlecht," *Briefe an Natalie,* 21.

13. Current descriptions of the gender balance in speech therapy in particular stress the ongoing scarcity of male speech therapists, e.g., S. Boyd and N. Hewlett, "The Gender Imbalance Among Speech and Language Therapists and Students," *International Journal of Language and Communication Disorders,* 36 (2001), 167–172.

14. Emma Seiler, *The Voice in Singing,* Philadelphia: J. B. Lippincott, 1893; Mrs. Emil Behnke [Kate Behnke], *The Speaking Voice, Its Development and Preservation,* London: Curwen & Sons, 1898.

15. *Briefe an Natalie,* 9.

16. Joseph Mainzer, *Méthode de chant pour les enfants à l'usage des écoles et des pensions,* vii.

17. In Germany, Johann Christoph Adelung published his influential dictionary of the German language in 1781 and the brothers Grimm initiated their dictionary project in 1852. The birth of a culturally "national" language is usually dated around the end of the eighteenth century. Peter von Polenz, *Deutsche Sprachgeschichte vom Spätmittelalter bis zur Gegenwart,* Berlin: Degruyter, 2000, 159. In France, the Académie Française was founded in the seventeenth century, but the centrifugal force of Paris would become particularly strong after the revolution, after which pronunciation was increasingly equalized as well. Arlette Farge, *Essai pour une histoire des voix au 18ième siècle.*

18. Theodor Siebs's manual for German pronunciation in the theater and Julius Hey's *Art of Speech* (now still in circulation and generally known as *Der kleine Hey*) codified the pronunciation of Hochdeutsch around 1900, but throughout the nineteenth century numerous authors on speech pathology suggested that different dialects (and particularly *Sächsisch*) were in fact speech impediments. Theodor Siebs, *Deutsche Bühnenausspache,* Bonn: Verlag Albert Ahn, 1898; Julius Hey, *Die Kunst der Sprache,* 1914.

19. Emile Sutro, *Duality of Thought and Language: An Outline of Original Research,* London: Dryden House, 1904, 78.

CHAPTER 3

Epigraph: Emil Behnke, *The Mechanism of the Human Voice,* London: J. Curwen and Sons, 1882.

1. Lennox Browne and Emil Behnke, *Voice, Song and Speech: A Practical Guide for Singers and Speakers, from the Combined View of Vocal Surgeon and Voice Trainer. With Numerous Illustrations by Wood-Engraving and Photography,* London: Sampson Low, Marston, Searle, and Rivington, 1884.

2. This is, after all, the time of the "gentleman scientist." On the intricacies of the private and public importance of nineteenth-century scientific work by gentlemen, see Rob Boddice, "Vivisecting Major: A Victorian Gentleman Scientist Defends Animal Experimentation, 1876–1885," *Isis,* 102:2 (2011), 215–237.

3. Behnke and Browne, *Voice, Song and Speech,* 5.

4. Castratos had been sought after as teachers of canto during their operatic heyday, but their influence extended far beyond that period. The continued reverence for the teachings of "legendary" masters such as Vaccai and Porpora was articulated in dynastic approaches to schools of singing in the nineteenth century, and cemented the notion that the Italian school of vocal education was superior to others. On the *bel canto* tradition and the place of the castrato within it, see John Potter, "The Tenor–Castrato Connection, 1760–1860." *Early Music,* 35 (2007), 97–112, and Brianna E. Robertson-Kirkland, "The Silencing of Bel Canto," *eSharp,* 21 (2013).

5. Special issue, "Scientific Masculinities," ed. Erika Lorraine Milam and Robert Nye, *Osiris,* 30/1 (2015).

6. On narratives of women's and children's "leaky" bodies and their impact on feminine and infantile voices, see, e.g., Gina Bloom, *Voice in Motion.*

7. W. Hayle Walshe, *Dramatic Singing, Physiologically Estimated,* London: Kegan Paul, 1881; Wesley Mills, *Voice Production in Singing and Speaking, Based on Scientific Principles,* London: Curwen & Sons, 1906; William H. Lawton, *The Singing Voice and Its Practical Cultivation,* New York: Charles Francis Press, 1901.

8. Emma Seiler, *The Voice in Singing,* 85.

9. William Shakespeare, *The Art of Singing,* 20.

10. Leo Kofler, *The Art of Breathing as the Bases of Tone-Production,* 23–24. Kofler's study started with a long autobiographical chapter charting the author's early career as a singer in church.

11. "sie machen es, ohne viel zu wissen, wie sie es machen." Johannes Müller, *Über die Compensation der physischen Kräfte am menschlichen Stimmorgan. Mit Bemerkungen über die Stimme der Säugethiere, Vögel und Amphibien,* Berlin: A. Hirschwald, 1839, i.

12. Behnke and Browne, *The Child's Voice: Its Treatment with Regard to After Development,* Chicago: Marquis and Co., 1885.

13. Alexander Melville Bell's *Lectures upon the Mechanism of Speech. Reprinted from the Proceedings of the First Summer Meeting of the American Association to Promote the Teaching of Speech to the Deaf* (New York: Funk and Wagnalls, 1906) quite literally reproduced Bell's lectures; Leo Kofler's *The Art of Breathing* referred to past lectures "before a society interested in breathing and singing"; Lennox Browne's *Science and Singing* (London: Chappell and Co., 1884) was based on shorthand reports of a public lecture. F. W. Mott's *The Brain and the Voice in Speech and Song* (London: Harper and Brothers, 1910) was likewise based on a series of lectures at King's College.

14. Cyril Rootham, *Voice Training for Choirs and Schools,* Cambridge, UK: Cambridge University Press, 1912, xxix.

15. Ibid.

16. Walter A. Yearsley, *A Practical Self-Cure of Stammering and Stuttering,* Copyrighted and Registered, published by Mrs. Yearsley, 1909, 17.

17. "(Ich bitte den Leser, sofort zu versuchen, 1. Durch Ohrenprobe, dann durch Kehlkopfprobe die reinen und stimmhaften Geräuschlaute zu bestimmen. Die Prüfung hat sich zu erstrecken auf die Konsonanten, die durch folgende Buchstaben wiedergegeben werden: f,h, l,m,n,r,s,ch, sch, w,b,p,g,k,d,t. Die Ergebnisse sind mit den Lauttafeln zu vergleichen; Dort sind die stimmhaften Geräuschlaute rot, die reinen schwarz gedruck. Alle stimmhaften Geräuschlaute sind singbar)." Alfred Hoffmann, *Was muss der Erzieher vom Wesen und von der Heilung der Sprach- und*

Stimmkrankheiten wissen? Zugleich eine Anleitung zum Studium der elementaren Lautkunde und zur Erlernung des richtigen Atmens, Meissen: Sächsische Schulbuchhandlung, 1910, 10.

18. For the seventeenth-century scientist, mind and body, or reason and affect, could be very closely connected and both could be equally important to the practices of science—even in an age generally characterized by formalism and Cartesian dualism. See, e.g., Nicholas D. Paige, *Being Interior: Autobiography and the Contradiction of Modernity in Seventeenth-Century France,* Philadelphia: University of Pennsylvania Press, 2000, 195–198, and Bernadette Höfer, *Psychosomatic Disorders in Seventeenth-Century French Literature,* Aldershot: Ashgate, 2009, 90–91.

19. This view was most famously articulated by the notion of the scientist as a modest witness in Steven Shapin and Simon Schaffer, *Leviathan and the Air-Pump: Hobbes, Boyle, and the Experimental Life,* Princeton, NJ: Princeton University Press, 1985 (and equally famously critiqued and modified in Donna J. Haraway, *Modest_Witness@Second_Millennium. FemaleMan©_Meets _OncoMouse*[a]*: Feminism and Technoscience,* New York: Routledge, 1997).

20. "Pendant longtemps l'oreille a été le seul moyen d'analyse; en effet, la photographie n'était pas suffisamment perfectionnée, on était obligé de faire soi-même les plaques; elles demandaient un certain temps pour être impressionnées, et l'on était réduit à étudier les vibrations avec l'oreille. Tout le monde n'a pas l'oreille aussi bonne.... On dit souvent qu'il ne faut discuter ni des goûts, ni des couleurs; mais discuter des sons est encore bien plus difficile." Mad. Marage, *Petit manuel de physiologie de la voix à l'usage des chanteurs et des orateurs,* Paris: chez l'auteur, 1911, 75–76.

21. As Alexandra Supper and Karin Bijsterveld have shown, attending to sound was increasingly professionalized in the nineteenth and twentieth centuries, and modern sonic skills included both modes of listening and using such new machines. Alexandra Supper and Karin Bijsterveld, "Sounds Convincing: Modes of Listening and Sonic Skills in Knowledge Making," *Interdisciplinary Science Reviews,* 40 (2015), 124–144.

22. "Nicht Neues zu geben war ihre Aufgabe, sondern Altes und Bekanntes einem grösseren Kreise vertraut zu machen." Paul Henry Gerber, *Die menschliche Stimme und ihre Hygiene: Sieben volkstümliche Vorlesungen*, Leipzig: G. B. Teubner, 1911, iii.

23. Marage, *Petit manuel.* The "performative" quality of scientific and medical endeavors had a long history, stretching back to the anatomical theaters of the Renaissance. By the nineteenth century, the almost theatrical scientific lecture had become so entrenched that the lectures in front of learned societies were heavily publicized (and published), and special lecture theaters were constructed for them. See, e.g., Frank A. J. L. James, "'Never Talk About Science, Show It to Them': The Lecture Theatre of the Royal Institution," *Interdisciplinary Science Reviews,* 27:3 (2002), 225–229.

24. Marage, *Petit manuel,* 89, 93.

25. E.g., Browne, *Science and Singing,* 10.

26. Browne, *Science and Singing,* 30.

27. David Knight, "Scientific Lectures: A History of Performance," *Interdisciplinary Science Reviews,* 27:3 (2002), 217–224.

28. Charles Bataille, *Nouvelles recherches sur la phonation: Mémoire présenté et lu à l'académie des sciences le 15 avril 1861,* p.v. "J'ai beaucoup expérimenté sur moi-même, et j'ai répété mes expériences sur des élèves familiarisés avec la méthode que je professe."

29. His *Reminiscences of a Stammerer* listed and reproduced a handful of lectures held before "large audiences." It is unclear who attended these lectures precisely, but the reactions recorded at one lecture organized by the "Literary and Scientific Institution" of Huntingdon suggests a gathering of generally "learned men" with an interest in scientific endeavor and its value for policymaking. Beasley, *Reminiscences of a Stammerer,* 129–131.

30. Beasley, *Reminiscences,* 139.

31. From *Birmingham Daily Gazette,* July 9, 1883, Benjamin Beasley, *Stammering: Its Treatment,* London: Waterlow and Sons Limited, 1888, 56.

32. Marchesi's *Vocal Method* became a manual for vocal teachers and students alike. The "manuals" of Caruso and Sims bore closer similarities to autobiographies than actual teaching materials. Enrico Caruso, *How to Sing? Some Practical Hints,* London: John Church Company, 1914; *Sims Reeves on the Art of Singing,* London: Chappell & Co., s.d.

33. Kofler, *Art of Breathing,* 6.

34. Caruso, *How to Sing,* 35.

35. Behnke and Browne, *The Child's Voice.*

36. Henry Holbrook Curtis, *Voice Building and Tone Placing: Showing a New Method of Relieving Injured Vocal Cords by Tone Exercises,* New York: D. Appleton and Company, 1909.

37. Henry Wood, *To My Pupils: Brief Notes to Be Kept in Memory, for Private Circulation Only,* 1896, 3.

38. Schmauk, *The Voice in Speech and Song,* 93.

39. Schmauk, *The Voice in Speech and Song,* 102.

40. Leo Klindworth, *Stuttering and How to Cure It.*

41. On the rise of ideas around universal, biology-based, notions of experience concomitant with the rise of modern science and comparative biology in particular, see, e.g., Rob Boddice and Mark Smith, *Emotion, Sense, Experience,* Cambridge, UK: Cambridge University Press, 2020.

42. Heinrich Pestalozzi, *Zur Vermeidung von Irrwegen bei der Ausbildung der menschlichen Stimme durch Indivuelle Stimmbildung,* Berlin: Albert Stahl, 1910, 9. "Gerade die Gegensätze der Theorien hatte ich Gelegenheit am eigenen Hals auszuprobieren. Ich sang oft lange Zeit auf diese oder jene Art, bis mich schlechte Erfahrungen dazu trieben, wieder etwas anderes zu versuchen. So sind mir Normalgesetze langsam klar geworden, und ich lernte das Richtige aus den verschiedenen 'Methoden' herausschälen. Darum bin ich auch überzeugt, dass jeder ehrlich strebende und forschende Mensch—natürlich auf ganz individuellem Wege—zu denselben Resultaten kommen wird."

43. Pestalozzi, *Zur Vermeidung von Irrwegen,* 19–21.

44. "Kolossale Stimmittel. Ihre Stimme tönt sehr voluminös aber stumpf."

45. Johannes Müller, *The Physiology of the Senses, Voice, and Muscular Motion, with the Mental Faculties,* London: Taylor, Walton, and Maberly, 1848, 1030.

46. F. W. Mott, *The Brain and the Voice in Speech and Song,* 34.

47. Henry Holbrook Curtis, *Voice Building and Tone Placing,* 165.

48. James Farrar, *The Human Voice and Connected Parts,* 54.

49. Müller, *The Physiology of the Senses,* 1030.

50. e.g., Holbrook Curtis, *Voice Building and Tone Placing.,* 168; W. A. Aikin, *The Voice: An Introduction to Practical Phonology,* London: Longmans Green, 1910, 99 and 140.

51. George Duncan Gibb, *On Diseases of the Throat and Windpipe, as Reflected by the Laryngoscope: A Complete Manual upon Their Diagnosis and Treatment. Embellished with 116 Engravings,* London: John Churchill, 1864 (2nd ed), viii.

52. Rather curiously, some singing teachers conversely claimed that experience with the laryngoscope was necessary for their craft too (Kofler, *The Art of Breathing,* 23). Morrell Mackenzie vigorously disagreed (Mackenzie, *Hygiene of the Vocal Organs,* 1886).

53. Hoffmann, *Krankheiten der Sprache und Stimme,* 10.

54. On the history of vibration as part of the history of sound, see Shelley Trower, *Senses of Vibration: A History of the Pleasure and Pain of Sound, London:* Continuum, 2012.

55. Wesley Mills, *Voice Production in Singing and Speaking Based on Scientific Principles,* 99.

56. George Gibb, *On Diseases of the Throat, Epiglottis and Windpipe,* 59.

57. In 1902, Arthur Chervin still mentioned all these inventions (ascribing many of them to Colombat), but essentially rejected them as useless. Chervin, *Bégaiement et autres maladies fonctionnelles de la parole,* 386.

58. Walter Yearsley, *A Practical Self-Cure of Stammering and Stuttering,* 148.

59. Beasley, *Stammering,* 21.

60. Kofler, *The Art of Breathing,* 17.

61. Lennox Browne, *Science and Singing,* 15.

62. Emil Behnke, *The Mechanism of the Human Voice,* London: J. Curwen and Sons, 1882, 15–16.

63. Schmauk, *The Voice in Speech and Song,* 59.

64. Behnke, *Mechanism of the Human Voice,* 25.

65. See, e.g., David Chapman, *Sandow the Magnificent: Eugen Sandow and the Beginnings of Bodybuilding,* Champaign: University of Illinois Press, 2006.

66. Richard B. Faulkner, *The Tonsils and the Voice in Science, Surgery, Speech and Song,* Pittsburgh: Presbyterian Book Store, 1913, 219–220.

67. Samuel Smiles would become famous—and infamous—as a proponent of a particularly individualistic Victorian approach to self-reliance in the second half of the nineteenth century, but methods to cure one's own ailments had been circulating long before that. John Hunter, *The Spirit of Self-Help: A Life of Samuel Smiles,* London: Shepheard-Walwyn, 2017.

68. As has been pointed out by historians of masculinity, the image of the self-made man that arose in the Victorian period was marked by inner conflicts between entrepreneurial and latent aristocratic ideals. E.g., David Kuchta, "The Making of the Self-Made Man," in *The Three-Piece Suit and Modern Masculinity: England, 1550–1850,* Oakland: University of California Press, 2002.

69. Historians of early modern Europe have, rightly, drawn attention to the importance of patients' corporeal experience for the development of early modern medical knowledge. This shifted for most disciplines in the late eighteenth century toward a more distanced approach with ambitions of objectivity, but the study of voice presents something of an exception to this overall development. See, e.g., Barbara Duden, *Geschichte unter der Haut: ein Eisenacher Arzt und seine Patientinnen um 1730,* Eberbach: Greif-Bücher, 1987.

70. T. Bartlett, *Stammering Practically Considered with the Treatment in Detail,* 11.

71. Schmauk, *The Voice in Speech and Song,* 94–95.

72. Kurt Schilling, *Schulpraktische Richtlinien für Singen, Sprechen und Rechtschreiben,* Meissen: Sâchsische Schulbuchhandlung, 1915, 9. "Schone Deine Stimme, Du klagst so fortwährend, und lass die Kinder vorsingen! Dieser Ton liegt der Auffassung der Kinder viel besser als Dein tiefer Basston, der sie nur zum Drücken verführt oder Deine mit grossem Kraftaufwand gebildete Höhe, die die Kinder wundervoll kopieren."

73. Cyril Bradley Rootham, *Voice Training for Choirs and Schools,* Cambridge, UK: Cambridge University Press, 1912, ix.

74. Schmauk, *The Voice in Speech and Song,* 97.

75. Both Theodore Schmauk and Morrell Mackenzie cited the *Stages of Men* in their handbooks. Theodore E. Schmauk, *The Voice in Speech and Song,* 97; Morell Mackenzie, *Hygiene of the Vocal Organs,* 52.

76. Josephine Hoegaerts, "Acquiring a 'Manly' Tone."

77. Frederick Helmore, *Speakers, Singers and Stammerers,* London: Joseph Masters and Co., 1874, 50.

78. Behnke and Browne, *The Child's Voice,* 80.

79. Caruso, *How to Sing*, 9.

80. Helmore, *Speakers, Singers and Stammerers*, 27.

81. Henry Wood, *The Gentle Art of Singing: In Four Volumes*, London: Humphrey Milford, 1927, 14.

82. Leopoldine Ziska, *Handbook of the Four Elements of Vocalization*. Written and dedicated to Lady Kerrison, London: Ewer & Co., 1850 (edited by Leopold Wray), 17.

83. Kofler, *The Art of Breathing*, 118.

84. Caruso, *How to Sing*, 26–27.

85. Archives Nationales, Pierrefitte, AJ 37/84, 192, 196, 215.

86. James Walker Downie, *Clinical Manual for the Study of Diseases of the Throat. With 104 illustrations*, Glasgow: James Maclehose and Sons, 1909, 394.

87. Denyse Rockey, *Speech Disorder in Nineteenth Century Britain.*

88. Benjamin Beasley, *Stammering: Its Treatment*, 55.

89. W. J. Ketley, Stammering: *The Beasley Treatment*, Birmingham: Hudson and Son, 1900, 74.

90. James Mackness, *Dysphonia Clericorum.*

91. John Bishop, *On Articulate Sound; and on the Causes and Cure of Impediments of Speech*, London: Samuel Highley, 1860, 46.

92. Bishop, *On Articulate Sound*, 78–79.

INTERMEZZO 3

1. W. J. Ketley, *Stammering: The Beasley Treatment*, 99–100.

2. Beasley, *Stammering: Its Treatment*, 24.

3. Beasley, *Reminiscences of a Stammerer*, 24.

4. Ibid., 42.

5. Ibid., 25.

6. Ibid., 54–56.

7. Beasley, *Stammering: Its Treatment*, 26.

8. Ibid., 27.

9. Ibid., 39.

10. Ibid., i.

11. Ibid., 13.

12. Ibid., 10.

13. Beasley, *Reminiscences*, 14.

14. *Birmingham Daily Gazette*, July 9, 1883, cited in Beasley, *Stammering*, 57.

CHAPTER 4

1. Emil Sutro, *Duality of Thought and Language*, 9.

2. Sutro, *Duality of Thought and Language*, 198.

3. This is, of course, not to say that laboratory research was new in the nineteenth century, but rather that the "modern" laboratory had a profound impact on how science was imagined in this period. As Robert E. Kohler has pointed out in "Lab History: Reflections," *Isis*, 99:4 (2008): "A history of the modern lab . . . would have to take into account the other essential, defining elements of the modernist package. These elements . . . include the scientific career and the scientific discipline,

with their elaborate social and moral infrastructures; and the bundle of reconfigured cultural concepts that includes objectivity, discovery, scientific method, and the social role of 'scientist.'"

4. Stephen M. Stigler describes the struggle, throughout the nineteenth century, to apply the then quite new statistical method to genetics and the social sciences (or, more prosaically, to come to an understanding of the "average man"). Stigler, *The History of Statistics: The Measurement of Uncertainty Before 1900*, Cambridge, MA: Belknap Press, 1986.

5. For early modern imaginations of the voice, see Gina Bloom, *Voice in Motion.*

6. As Jonathan Sterne has pointed out, the possibility of playback was central to the process of inventing recording technology—even before the requisite technology was available: "Without the possibility of playback, what good is the recording of sound? In this way, we might say that, insofar as sound technologies are *ever* organized into sound media, the medium—or, at least, an imagined medium—precedes even the technology itself." Sterne, *The Audible Past*, 214.

7. John Picker, "The Recorded Voice from Victorian Aura to Modernist Echo," 110–146.

8. See, e.g., Joshua St. Pierre and Charis St. Pierre, "Governing the Voice: A Critical History of Speech-Language Pathology," *Foucault Studies*, 24 (2018, 151–184.

9. Alfred Appelt, *The Real Cause of Stammering and Its Permanent Cure.*

10. L. Morin, *Traité de prononciation, indiquant les moyens d'obtenir une bonne émission de voix, de corriger les accents vicieux et tous les défauts de prononciation*, Paris: Passage Saulnier, 1852. "Vous croyez qu'un fou peut vivre sans cervelle?/ Erreur, erreur; le mal n'est point dans le cerveau;/ Il est . . . dans les poumons . . . n'est-ce pas? C'est nouveau."

11. Honoré Mathieu, *De la parole et du bégaiement*, Paris: Chez l'auteur, 1847. "L'esprit est l'ensemble de toutes les facultés qui ont leur siège dans le cerveau, telles que l'imagination, l'intelligence, la mémoire, etc. Ainsi, soit que les idées émanent de nous, soit qu'elles nous viennent d'autrui, dans l'esprit."

12. Henry Monro, *On Stammering and Its Treatment.*

13. J. Duquesnois, *Manuel de l'orateur et du lecteur.* "Je crois, messieurs, que vous admettrez sans peine avec moi que la voix est pour les oreilles ce que l'écriture est pour les yeux, c'est-à-dire qu'elles doivent l'une et l'autre, par des routes différentes, il est vrai, transmettre les pensées au cerveau."

14. Vernet, *Leçons théoriques et pratiques sur la cause, le siège, la nature, le mécanisme et le traitement du bégaiement, ou Méthode naturelle de traitement du bégaiement en 20 leçons, mise à la portée de tout le monde par Mlle Coralie Vernet et Claude Vernet son père.*

15. Louis Profumo, *De l'organe de la parole interne*, Paris: Imprimerie de Pillet, 1860. "On n'ignorait pas en physiologie cérébrale que le siège de la pensée était dans la partie antérieure et supérieure du cerveau; c'est même le seul fait clair et sùr dans l'obscurité et le doute qui font de la physiologie psychologique du cerveau une science au moins bien problématique."

16. Ibid. "Le but de mon travail, c'est de démontrer l'existence dans la moitié antérieure du cerveau d'un organe de la parole interne."

17. Ibid. "Mais la parole de la bouche se fait entendre aux autres, tandis que la parole du cerveau n'est aperçue que seulement par le même individu qui la produit, et même dans l'état normal elle fait impression sur le moi sans qu'il en ait conscience."

18. Ibid. "certainement dans la partie antérieure et supérieure du cerveau."

19. Broca, P. "Du siège de la faculté du langage articulé dans l'hémisphère gauche du cerveau," *Bulletins de la Société d'anthropologie*, IV (1865), 8: "les actes beaucoup plus simples et beaucoup plus grossiers."

20. Ibid., "l'écriture, le dessin, la broderie, etc."

21. Ibid., 16. "le développement plus précoce de l'hémisphère gauche nous predispose, dans nos premiers tâtonnements, à executer avec cette moitié du cerveau les actes matériels et intellectuels les

plus conpliquées, parmi lesquels il faut certainement competer l'expression des idées au moyen du langage et, plus particulièrement, du langage articulé."

22. Ibid., 4. "Certes, l'observation est supérieure aux theories, et il faut savoir quelquefois s'incliner devant un fait, quelque inexplicable, quelque paradoxal qu'il puisse nous paraître."

23. Broca's influence on the study of aphasia, in particular, is beautifully laid out in Heini Hakosalo's history of the "discovery" of aphasia and the formation of the "field." Heini Hakosalo, *On Speaking Terms: Scientific Boundary Work and the Discovery of Aphasia, 1861–1874*, Oulu: Doctoral dissertation, 2006, 15–24.

24. Broca, "Du siège de la faculté du langage articulé," 3. "Je n'aime pas les discussions de priorité."

25. Marc Dax, "Lésion de la moitié gauche de l'encéphale coïncidant avec l'oubli des signes de la pensée," *Gazette hebdomadair de medicine et de chirurgie*, 1865.

26. Adolf Kussmaul, *Die Störungen der Sprache: Versuch einer Pathologie der Sprache*, Handbuch der speciellen Pathologie und Therapie, XII, Leipzig: F.C.W. Vogel, 1877, 135. "Der Berichterstatter Lélut ging mit wenigen schnöde abweisenden Worten kurz darüber weg und erklärte alle Versuch, die Sprache zu localisieren, schlechthin für Phrenologie."

27. Albert Carrier, *Étude sur la localisation de la faculté du langage articulé*, Paris: Germer Baillière, 1897.

28. Carrier, *Étude sur la localisation de la faculté du langage*, 7.

29. Ibid., 8. "regardée comme une utopie par les uns, comme une hallucination par les autres, elle n'avait jamais été formulae d'une façon rigoureuse, et était tombée dans l'oubli."

30. Ibid. "donna à ce principe les localisations toute la netteté et l'importance qu'il méritait."

31. Adolf Kussmaul, *Versuch einer Pathologie der Sprache*, Leipzig: F.C.W. Vogel, 1885. Kussmaul is credited with being the first to describe, among other things, selective mutism and word blindness (now commonly known as dyslexia).

32. Abraham Netter, *La parole intérieure et l'âme*, Paris: Berger-Levrault, 1892, 28

33. F. W. Mott, *The Brain and the Voice in Speech and Song*, 4.

34. Marjorie Lorch, "The Long View of Language Localization," *Frontiers in Neuroanatomy*, 13:52 (2019).

35. Joseph Collins, *The Genesis and Dissolution of the Faculty of Speech*, 2.

36. Ibid., 40.

37. F. W. Mott, *The Brain and the Voice*, 103.

38. Albert Lemoine, *De la physionomie et de la parole*, 14. "le problème de l'origine du langage est unsoluble comme problème historique," "une question de psychologie."

39. Albert Deleschamps, *Etude physique des sons de la parole*, Paris: Savy, 1869, 59. "Il est formé par l'alliance indissoluble d'un son et d'un sens, d'un élément phonétique et d'un élément psychique, la pensée et l'expression, la chose signifiée et le signe."

40. Arthur Chervin, *Bégaiement et autres maladies fonctionnelles de la parole*, Paris: Société d'éditions scientifiques, 1902, 158.

41. Chervin, *Bégaiement et autres maladies*, 158–159. "Je citerai le fait d'un officier atteint d'un bégaiement assez léger relativement et qui un jour, pendant qu'il faisait exécuter des exercices de marche à ses soldats, se faisait ce raisonnement: «Si, par hasard, je n'allais pas pouvoir les arrêter!» Les militaires marchaient au pas, et l'officier qu'il était toujours préoccupé de cette idée qu'il pourrait, peut-être, ne pas pouvoir leur crier: halte! Cette réflexion l'obsédait. Lorsqu'il voulut faire le commandement, il ne le put pas . . . et les soldats, qui s'exerçaient dans un terrain borne par un fossé, avançaient toujours. L'officier les voyait bien se diriger et se rapprocher davantage de ce fossé, mais il

se sentait absolument incapable de leur crier: halte! Enfin, les voyant sur le bord du fossé, il fit un violent effort sur lui-même et poussa un cri quelconque. Comme les soldats ne demandaient pas mieux que de s'arrêter,' ils comprirent et s'arrêtèrent à la grande satisfaction de leur malheureux officier."

42. Alfred Appelt, *The Real Cause of Stammering and Its Permanent Cure,* 41–42.

43. Ibid., 43.

44. Ibid., 132, 143.

45. Richard Lynch Garner, *The Speech of Monkeys,* London: William Heinemann, 1892, 170.

46. Nottidge C. Macnamara, *Human Speech: A Study in the Purposive Action of Living Matter,* London: Kegan Paul, Trench & Co, 1908, 192.

47. F. W. Mott, *The Brain and the Voice in Speech and Song,* New York: Harper & Brothers, 1910, 4.

48. Albert Lemoine, *De la physionomie,* 12. "Par exemple chaque enfant nouveau-né ne crée pas de toutes pièces la langue de son pays, mais on ignore, parce qu'on ne l'observe pas d'assez près, tout le génie que dépense cette petite créature durant son instruction première."

49. On the overlap between the medical and colonial gaze toward sign language (and the communities and individuals using them), see, e.g., Esme Cleall, "Deafness: Representation, Sign Language, and Community, ca. 1800–1920," in J. L. Huff and M. Stoddard Holmes (eds.), *A Cultural History of Disability in the Long Nineteenth Century*. London: Bloomsbury, 2020, 95–112. More generally, philology played a large role in developing ethnic as well as national categories of "language families" using evolutionary discourses; see, e.g., Will Abberley, "Race and Species Essentialism in Nineteenth-Century Philology," *Critical Quarterly,* 3:4 (2011).

50. These interpretations also drew on older theories about the sounds of language and culture, as developed for example by J. G. Herder in the eighteenth century. Tanvi Solanki, "Listening to Difference: J. G. Herder's Aural Theory of Cultural Diversity in the 'Treatise on the Origin of Language' (1772)," 930–947.

51. In a recent memoir, for example, talking is credited with the ability to further civility: Paula Marantz Cohen, *Talking Cure: An Essay on the Civilizing Power of Conversation,* Princeton, NJ: Princeton University Press, 2023.

52. Lewis B. Monroe, *Manual of Physical and Vocal Training, for the Use of Schools and for Private Instruction,* Philadelphia: Cowperthwait, 1869, 69.

53. Susan Broomhall, *Early Modern Emotions: An Introduction,* New York: Routledge, 2017.

54. Theodore Schmauk, *The Voice in Speech and Song,* 120.

55. Ibid., 122.

56. Josephine Hoegaerts, *Speaking Like Intelligent Men*, 123–144.

57. Helmore, *Speakers, Singers and Stammerers*, 51.

58. See, e.g., Sarah Richardson, *The Political Worlds of Women: Gender and Politics in 19th-Century Britain,* New York: Routledge, 2013.

59. Shakespeare, *The Art of Singing,* 50.

60. The instrument was advertised by a substantial "jingle," spread as sheet music to be performed at home. Percy G. Mocatta and Alfred Allen, *The Lost Voice (A Song of the Ammoniaphone),* London: Medial Battery Company, 1880.

61. As Allan Reddick has noted, for example, the standardization of pronunciation in the eighteenth and nineteenth centuries was often driven by a fear of improper sounds like those of Irish or Scottish. Reddick, "Recording the Most Proper and Significant Words," in Sarah Ogilvie (ed.), *Eighteenth-Century English Dictionaries: Prescriptivism and Completeness,* Cambridge, UK: Cambridge University Press, 2020.

62. The work was published not long after Duden published its first handbook of grammatical and orthographic rules of German, in 1880. Julius Hey, *Deutsches Gesangsunterricht,* 1900.

63. This representation of dialects as both impure and "out of place" in the national context shares much of its vocabulary with the language used to describe immigrants: using metaphors of invasion to describe an entangled issue of health and citizenship. See, e.g., Philip Sarasin, *Anthrax. Bioterror als Phantasma,* Frankfurt a.M: Suhrkamp, 2004; Nirmal Puwar, *Space Invaders: Race, Gender and Bodies Out of Place,* Oxford: Berg Publishers, 2004.

64. Brunot took a scientific interest in phonology and pronunciation, using the newly invented phonograph to document different types of speech. The Archives de la Parole contain numerous instances of both standardized and non-standard speech, and they are labelled as such. "La parole au timbre juste, au rythme impeccable, à l'accent pur" and "la parole nuancée d'accents faubourien ou provincial."

65. On the development of a genealogical model of language in nineteenth century linguistics, see Stephen G. Alter, *Darwinism and the Linguistic Image: Language, Race and Natural Theology in the Nineteenth Century,* Baltimore: Johns Hopkins University Press, 2003.

66. Ironically, this narrative of a naturally brilliant voice was placed at the beginning of a manual to teach readers how to sing. Caruso, *How to Sing.*

67. Taru Haapala, *Political Rhetoric in the Oxford and Cambridge Unions, 1830–1870,* London: Palgrave Macmillan, 2016.

68. Lucy Green, *Music, Gender and Education,* Cambridge, UK: Cambridge University Press, 1997.

69. Leo Kofler, *The Art of Breathing as the Basis of Tone-Production,* 104.

70. On the gradual professionalization of choral societies and their social function, see, e.g., Guillaume Lurton, *Le Chœur partagé. Le chant choral en France, intégration socio-économique d'un monde de l'art moyen,* doctoral thesis, Sociologie. Sciences Po—Institut d'études politiques de Paris, 2011.

71. See, e.g., J. Varley Roberts, *A Treatise on a Practical Method of Training Choristers,* London: Oxford University Press, 1900, 13.

72. Josephine Hoegaerts, "Little Citizens and Petites Patries: Learning Patriotism Through Choral Singing in Antwerp in the Late Nineteenth Century," in U. Geisler and K. Johansson (eds.), *Choral Singing: Histories and Practices,* Cambridge, UK: Cambridge Scholars Publishing, 2014, 14–32.

73. Roberts, *A Treatise on a Practical Method,* 11.

74. Dave Russell, *Popular Music in England 1840–1914: A Social History,* Manchester: Manchester University Press, 1997.

75. See, e.g., Svenja Goltermann, *Körper der Nation. Habitusformierung und die Politik des Turnens, 1860–1890,* Göttingen: Vandenhoeck & Ruprecht, 1998.

76. Julien Tiersot, *Programme d'un recueil de chants à l'usage des écoles primaires de France,* Paris: Librairie Hachette et cie, 1893, p. vi. "quel incomparable agent de civilisation morale et d'éducation nationale est le chant choral, quand le langage de la musique expriment, chacun à sa manière, des sentiments généraux, de ceux auxquels tout coeur d'homme répond."

77. Mackenzie, *The Hygiene of the Vocal Organs,* 171.

78. This equation between aesthetics, health, and morality was not unique to representations of the voice, but applied to the body and its visual characteristics as well. See, e.g., Lucy Hartley, "Constructing the Common Type: Physiognomic Norms and the Notion of 'Civic Usefulness', from Lavater to Galton," in Waltrand Ernst (ed.), *Histories of the Normal and the Abnormal: Social and Cultural Histories of Norms and Normativity,* New York: Routledge, 2006, 101–121.

79. Georg Hermann von Meyer, *The Organs of Speech and Their Application in the Formation of Articulate Sounds,* London: Kegan Paul, Trench & Co., 1883, 219.

80. See, for example, Ville Kivimäki, Tanja Vahtikari, and Sami Suodenjoki, *Lived Nation: Histories of Experience and Emotion in Understanding Nationalism,* London: Palgrave, 2020.

81. M. F. Boisquet, *Essais sur l'art du comédien chanteur,* Paris: Longchamps, 1812, 51. "On a remarqué que dans les pays chauds les voix et la prononciation étaient plus gutturales, que les articulations dépendaient plus précisément de l'action des parties antérieures de l'organe, au lieu que dans les pays froids toutes les articulations sont labiales et sont dues manifestement à l'action des parties antérieures de l'instrument vocal. On voit aisément la raison de cet effet. Une température douce invite la nature à dilater l'organe de la voix et à lui permettre de déployer toute l'étendue du mouvement dont il est susceptible, au lieu que l'impression du froid le retient dans un état de contrainte et de resserrement continuels; en sorte que chez les habitans des pays froids, c'est sure les parties les plus externes de l'organe que doivent porter tout l'effort et toute l'action de la voix; ce qui vient à l'appui de ce que j'avance."

82. S. Faure-Lacroze, *Essai sur la composition d'un nouvel alphabet pour servir à représenter les sons de la voix humaine,* Paris: Firmin Didot frères, fils et cie, 162–163. "les sons durs et les consonnes multipliées dans la même syllabe semblent annoncer des nations entêtées, pensantes, sérieuses et flegmatiques, comme les Allemands: des nations civilisées, obésiteuses et silencieuses comme les Anglais, les Hollandais, dans un climat froid, auront probablement des sons gênés et paresseux: des nations gaies, doucereuses et voluptueuses dans un climat riant, comme les Italiens, les Haïtitiens, auront dans leurs sons beaucoup de voix sonores et à formes riantes: des nations graves et furibondes, telles que les sauvages de l'Amérique, les Espagnols, les Turcs, etc., quel que soit d'ailleurs le climat, auront des sons pleins et graves; tandis que des peuples civilisés, humains, industrieux, et surtout causeurs comme les Français, dans un climat tempéré, doivent avoir un grand nombre de sons, mais ne doivent de sons durs et pas se servir fréquemment graves, non plus que de doubles et triples consonnes."

83. Alexander Melville Bell, *Visible Speech,* London: Simpkin, Marshall & Co., 1867, 21. This goal was swiftly followed by the addition that colonizers would of course also be able to learn local languages; but this endeavor was clearly connected to their need to "rule" effectively, or to classify, study, and protect local languages.

84. John Wyllie, *The Disorders of Speech,* Edinburgh: Oliver and Boyd, 1898, 155–156.

85. These parallels between the representation and approach to ethnic and Deaf communities have had an important influence on the development of Deaf studies as a field, as well. Modes of social justice and antiracist critique have helped to shape Deaf studies as a discipline in the twentieth century, alongside colonialist discourses—most obviously, perhaps, in the work of Harlan Lane. Esme Cleall, "Decolonising Deaf History. Harlan Lane, Postcolonialism, and Critical Colonial History," *Zeithistorische Forschungen/Studies in Contemporary History,* 19 (2022).

86. Wyllie, *Disorders of Speech,* 154–155.

87. Jennifer Lynn Stoever, *The Sonic Color Line.* On Black voices on stage in the nineteenth century, see, e.g., Naomi André, *Black Opera: History, Power, Engagement,* Champaign: University of Illinois Press, 2018; Kira Thurman, *Singing Like Germans: Black Musicians in the Land of Bach, Beethoven and Brahms,* Ithaca, NY: Cornell University Press, 2021.

88. The most famous example of such inabilities/refusals to understand ethnicized language is perhaps a trial over a Canadian land claim, Delgamuukw v. the Queen (1985). For an analysis of this trial as an instance of racialized listening, see Dylan Robinson, *Hungry Listening,* 41–46. David Le Breton's *Du Silence* likewise approaches the case as an instance of reducing speech to silence (Paris: Metallié, 1997).

89. Bourke, *What It Means to Be Human.*

90. On the connection between infantilization, silence, and lack of civil rights, see, e.g., Josephine Hoegaerts, "Recording the Subaltern's Speech. Children's Voices in the Antwerp School Archives, ca.1850–1900," *Journal of Belgian History,* 46:1 (2016), 10–31.

91. Richter, *Demokratie, eine Deutsche Affäre,* Munich: C.H. Beck, 2021, 22.

92. Bruce Thomas Boehrer, *Parrot Culture: Our 2500-Year-Long Fascination with the World's Most Talkative Bird,* Philadelphia: University of Pennsylvania Press, 2004.

93. Adolf Kussmaul, *Störungen der Sprache. Versuch einer Pathologie der Sprache,* 55.

94. Ibid.

95. Arthur Chervin, *Exercices de lecture à haute voix et de récitation, divisions élémentaires: prononciation française,* Paris: Chez l'auteur, 1880, 262. "Rien n'est fatigant, en société, comme les enfants babillards, qui interrompent à tout instant les conversations sérieuses. On les compare, a juste titre, à des perroquets, dont l'intarissable babil amuse un moment, mais finit bientôt par lasser et devenir insupportable."

96. Thomas Dixon, *The Invention of Altruism: Making Moral Meanings in Victorian Britain,* Oxford: Oxford University Press, 2008, 140.

97. Abraham Netter, *La parole intérieure et l'âme,* 123.

98. Ibid., 124. "Vous connaissez l'histoire du lapin de M. Laborde, histoire vraie, authentique: 'Le lapin dont il s'agit a été légendaire au vieux laboratoire de physiologie: toutes les personnes qui ont fréquenté cet établissement de 1877 à 1880 n'ont pas oublié l'illustre rongeur qui répondait au nom de "Bertrand".'"

99. Ibid., 127. "Il assistait comme s'il eût voulu y prendre part, et en tous cas, ayant l'air de s'y intéresser, à mes travaux d'expérimentation. L'expression et l'attitude de la curiosité se peignaient alors dans ses allures; il restait des heures entières sur la table d'expériences où était attaché le sujet, chien ou lapin, comme en observation avec nous. Mais ce qui a toujours paru l'intéresser ou l'intriguer le plus, c'est l'examen microscopique. A peine étais-je installé au microscope, que Bertrand s'empressait de monter sur la table ou sur mes épaules, et de se placer exactement dans la position qu'il me voyait prendre, collant son oeil à la lunette microscopique."

100. Rob Boddice, "Vivisecting Major: A Victorian Gentleman Scientist Defends Animal Experimentation, 1876–1885," 215–237, 223.

101. Ibid., 236.

102. Nottidge C. Macnamara, *Human Speech,* 176–193.

103. Ibid., 191.

104. Bruce Thomas Boehrer, *Parrot Culture,* 5.

105. N. Macnamara, *Human Speech,* 191.

106. Ibid., 185.

107. Ibid., 248–251. See also Joanna Bourke, *What It Means to Be Human,* 29–60.

108. Boehrer, *Parrot Culture,* 4–11.

109. Frederick W. Mott, *The Brain and the Voice in Speech and Song,* 4.

110. Ibid.

111. Mott, *The Brain and the Voice,* 7.

112. Ibid., 60.

113. Macnamara, *Human Speech,* 187.

114. Albert Lemoine, *De la physionomie et de la parole,* 188–189. "Que manque-t-il donc à l'homme pour qu'il soit condamné à un mutisme éternel, si quelque miracle ne s'opère au dedans ou au dehors de lui? Est-ce comme au poisson la voix, comme au singe la puissance d'articuler des sons, comme au perroquet la raison, où l'ouïe comme au sourd?"

115. Lemoine, *De la physionomie*, 179. "C'est l'idiot enfin que sa raison imbécile rapproche du perroquet bavard, ce n'est pas le sourd."

116. Gardiner, *Music of Nature*, 22.

INTERMEZZO 4

1. Karin Bijsterveld, *Sonic Skills: Listening for Knowledge in Science, Medicine and Engineering*, London: Palgrave Macmillan, 2019.

2. Jonathan Sterne, *MP3, the Meaning of a Format*, Durham, NC: Duke University Press, 2012.

3. Brian Kane, *Sound Unseen: Acousmatic Sound in Theory and Practice*, Oxford: Oxford University Press, 2014.

4. Anna Harris and Tom Rice, *Stethoscope: The Making of a Medical Icon*, London: Reaktion Books, 2022.

5. Joeri Bruyninckx, *Listening in the Field*.

6. Richard Lynch Garner, *The Speech of Monkeys*, 5.

7. Ibid., 1–2.

8. Ibid., 5.

9. Ibid., 6.

10. Ibid., 9.

11. Ibid., 34.

12. Ibid., 149.

13. Ibid., 96.

14. Ibid., 66–67.

15. Ibid., 35.

16. Ibid., 3.

17. See, e.g., Ingrid L. Nelson, "Interspecies Care and Aging in a Gorilla 2.0 World," *Geoforum*, 79 (2017), 144–152.

18. See also Joanna Bourke, *What It Means to Be Human*, Chapter 2.

19. Franz Kafka, "A Report for An Academy," translated by Ian Johnston. Original: Franz Kafka, "Ein Bericht für eine Akademie," *Der Jude. Eine Monatsschrift*. Jg. 2, Heft 8., 1917. "Es war so leicht, die Leute nachzuahmen. Spucken konnte ich schon den ersten Tag. Wir spuckten einander dann gegenseitig ins Gesicht; der Unterschied war nur, daß ich mein Gesicht nachher reinleckte, sie ihres nicht."

20. Garner, *The Speech of Monkeys*, 110.

21. Monica Mattfeld, *Becoming Centaur: Eighteenth-Century Masculinity and English Horsemanship*, University Park: Penn State University Press, 2016.

CHAPTER 5

1. It is also not entirely correct, and played a part in mythmaking endeavors around the telephone as a new medium. See, e.g., J. Sterne, *The Audible Past*, 31.

2. Sarah Jackson, "Derrida on the Line," *Derrida Today*, 10:2 (2017), 142–159; Daniel Martin, "Speaking Machines and Ghostly Phantoms: The Claustrum Poetics of Voice and Dysfluency," *Victorian Review*, 46:1 (2020), 1–5.

3. See, e.g., John Picker, *Victorian Soundscapes*, 110–133.

4. Anthony Enns, "The Human Telephone: Physiology, Neurology and Sound Technologies," in Daniel Morat (ed.), *Sounds of Modern History: Auditory Cultures in 19th- and 20th-Century Europe,* New York: Berghahn, 2014, 46–68.

5. *Atheneum,* July 15, 1865, reproduced in Alexander Melville Bell, *Visible Speech,* 29.

6. Pitman's *Shorthand Phonography,* London: Isaac Pitman and Sons, 1837, is the most obvious and most famous example, but other methods made similar claims, e.g., Hyppolyte Prévost, *Nouveau manuel de sténographie ou l'art de suivre la parole en écrivant,* Paris: Roret, 1834; Emile Duployé, *Sténographie Duployé, ou L'Art de suivre avec l'écriture la parole la plus rapide appris sans maître en quelques heures,* Aisne: Duployé, 1867.

7. Adelheid Voskuhl, *Androids in the Enlightenment: Mechanics, Artisans and Cultures of the Self,* Chicago: University of Chicago Press, 2013.

8. As, for example, Arthur Asseraf and Iwan Rhys Morus have shown, electricity served as an exciting explanatory metaphor and practice in the nineteenth century, used in contexts as disparate as news reporting and medical intervention. It would also grow to play a very important role in psychiatry. Arthur Asseraf, *Electric News in Colonial Algeria,* Oxford: Oxford University Press, 2019; Iwan Rhys Morus, *Shocking Bodies: Life, Death & Electricity in Victorian England,* Cheltenham: History Press, 2011.

9. Kratzenstein's vowel machine is often seen as an early attempt to synthesize speech. John J. Ohala, "Christian Gottlieb Kratzenstein: Pioneer in Speech Synthesis," *ICPhS,* XVII (2011), 156–159.

10. Madame Marage, *Petit manuel de physiologie de la voix à l'usage des chanteurs et des orateurs.*

11. Rob Mullender, "Divine Agency. Bringing to Light the Voice Figures of Margaret Watts-Hughes," *Sound Effects,* 8:1 (2019), 123–139.

12. Fay Bound Alberti, *Matters of the Heart: History, Medicine and Emotion,* Oxford: Oxford University Press, 2010.

13. John Picker, *Victorian Soundscapes,* 110–126.

14. *The Granta,* February 2, 1895, 169.

15. Kerim Yasar, *Electrified Voices. How the Telephone, Phonograph, and Radio Shaped Modern Japan, 1868–1945,* New York: Columbia University Press, 2018.

16. In that sense, the telephone can be counted among such recording devices, even though storages and repeatability were not among its functions.

17. Research on the catalogues of these early recording specialists has long focused on who or what was recorded, rather than the acoustic aspects of the recordings, as damage to cylinders and disks (as well as machinery becoming obsolete) has made the recordings unreadable. More recently, attempts have increasingly been made to recover the sounds on these recordings, thus allowing access to new data—but also further complicating the relationship between historians of sound and their primary sources as different forms of mediatization become intermingled. See, e.g., work on sound recordings in the Smithsonian collection at the National Museum of American History. https://americanhistory.si.edu/press/fact-sheets/early-sound-recording-collection-and-sound-recovery-project

18. Katherine Bergeron, *Voice Lessons: French Melodie in the Belle Epoque*; Freya Jarman, *Queer Voices*; Roland Barthes, "Le grain de la voix," *L'obvie et l'obtus,* Seuil, 1982.

19. See for example James Q. Davies, "Pneumotypes: Jean de Reszke's High Pianissimos and the Occult Sciences of Breathing."

20. Karin Martensen, *The Phonograph Is Not an Opera House.*

21. The importance of the role of mediatization in the production (and analysis) of paper documents has also been demonstrated by, e.g., Lisa Gitelman, *Paper Knowledge. Toward a Media History of Document,* Durham, NC: Duke University Press, 2014; Delphine Gardey, *Écrire, Calculer,*

Classer. Comment une révolution de papier a transformé les sociétés contemporaines (1800–1940), Paris: Éditions la découverte, 2008.

22. Walter Ong, "Writing Is a Technology That Restructures Thought," in *The Linguistics of Literacy*, ed. Pamela Downing, Susan Lima, Michael Noonan, Amsterdam: J. Benjamins, 1992, 293–319; Walter Ong, *Orality and Literacy: The Technologizing of the Word*, London: Methuen & Co., 1982.

23. Republished as Donna Haraway, "A Cyborg Manifesto: Science, Technology, and Socialist-Feminism in the Late Twentieth Century," in *Simians, Cyborgs and Women: The Reinvention of Nature*, New York: Routledge, 1991, 149–181.

24. As Holger Schulze has pointed out, this has also brought back a social practice of interacting with domestic servants—nonhuman, but often nevertheless gendered ones. E.g., in an interview between Schulze and Julia Hofer, "Unterwürfighe Sprachassistentinnen," *Beobachter Digital*, 4 June 2020, https://www.beobachter.ch/digital/unterwurfige-sprachassistentinnen-alexa-zementiert-geschlechterrollen-von-kindesbeinen.

25. Great strides have been made, for example, in embodied methods for fashion history, food history and of course music history. See, e.g., Hilary Davidson, "The Embodied Turn: Making and Remaking Dress as an Academic Practice," *Fashion Theory*, 23:3 (2019), 329–362.

26. Katherine Bergeron takes a similar approach in *Voice Lessons*.

27. Nicholas Cook, *Beyond the Score*.

28. Leonard Lawlor, "Auto-Affection," *Derrida: Key Concepts*, New York: Routledge, 2014, 130–138.

29. Martensen, *The Phonograph Is Not an Opera House*, 171–173.

30. James Q. Davies, "The Sontag-Malibran Stereotype," *Romantic Anatomies of Performance*, Oakland: University of California Press, 2014, 66–92.

31. Wayne Koestenbaum, *The Queen's Throat*, 49.

32. Steven Connor, *Dumbstruck. A Cultural History of Ventriloquism*, 3–4.

33. Martha Feldman, *The Castrato*.

34. Mladen Dolar has suggested that the speaking puppet is "the real puppet," the ghost in the machine—particularly of the modern voice. *A Voice and Nothing More*, 10.

35. James Q. Davies, "Pneumotypes," 23.

36. Ibid., 29.

37. Ibid., 43.

38. Breathing, like vocalizing, is a historically contingent practice, and in the (late) nineteenth century it became increasingly defined by notions of biology, chemistry, and psychology, leading to thinking about breathing as a human/animal "universal." Oriana Walker, *The Breathing Self: Toward a History of Respiration*. Doctoral dissertation, Harvard University, 2016.

39. On singing into the mask, its historical meaning, and continued use in bel canto technique, see, e.g., Daniel Shigo, *Voice Talk*, "The Garcia Method and Singing in the Mask," https://www.shigovoicelessons.com/voicetalk//2014/07/the-garcia-method-singing-in-mask.html.

40. James Q. Davies, "Pneumotypes," 23.

41. Ibid., 22.

CONCLUSION

1. See, e.g., Nina Sun Eidsheim, *The Race of Sound: Listening, Timbre and Vocality in African American Music*.

2. Bourke, *What It Means to Be Human*, 256–263.

3. See also Butler, *Grievable Lives*.

4. Bourke, *What It Means to Be Human*, 251.

5. Ibid.

6. The rise of neurology and psychology contributed to this shift, as shown in Chapter 4.

7. Joshua St. Pierre, *Cheap Talk*, 14.

8. Roland Barthes, *Image, Music, Text*, 269.

9. Barthes's distinction between pheno-song and geno-song implicates both the singer and the listener, distinguishing between a type of singing that is heard as mostly emotive and expressive (pheno-song), and a singing that centers the sounds and materiality of the singer's body (geno-song).

10. Ibid., 270.

11. Freya Jarman, *Queer Voices*.

12. Barthes, *Image, Music, Text*, 278.

13. Jarman, *Queer Voices*, 20.

BIBLIOGRAPHY

PRIMARY READING

Leipzig: Samuel Heinicke Library

Angermann, F. *Lautlehre mit einem Anhange: die Fehler der Aussprache, ausgenommen das Stottern. Für Elementar- und Gesanglehrer, Sprachärzte und Pädagogen.* Berlin: Verlag von Carl Reimarus, 1850.

Angermann, F. *Das Stottern, sein Wesen und seine Heilung. Für Ärzte und Pädagogen.* Berlin: Druck und Verlag von Trowitzsch und Sohn, 1853.

Aronsohn, Oscar, *Der psychologische Ursprung des Stotterns.* Halle a.S.: Carl Marhold Verlagsbuchhandlung, 1914.

Assmann, Julius, *Das Stottern, Ein Beitrag zum Verständnis und zur Heilung desselben. Für Lehrer und erwachsene Sprachleidende.* Hamburg and Berlin: Verlag von Bruer & Co., 1890.

Aubert, Hermann, Die innerliche Sprache und ihr Verhalten zu den Sinneswahrnehmungen und Bewegungen. *Zeitschrift Psychologische Und Physiologische Sinnesorg,* 1:1, 1890.

Ballet, Gilbert, *Die innerliche Sprache und die verschiedenen Formen der Aphasie.* Leipzig and Vienna: Franz Deuticke, 1890.

Barth, Adolf, *Über die Bildung der menschlichen Stimme und ihres Klanges beim Singen und Sprechen.* Leipzig: Verlag von Johann Ambrosius Barth, 1904.

Barth, Ernst, *Einführung in die Physiologie, Pathologie und Hygiene der menschlichen Stimme.* Leipzig: Verlag von Georg Thieme, 1911.

Bastian, Charlton, *Über Aphasie und andere Sprachstörungen.* Leipzig: Verlag von Wilhelm Engelmann, 1902.

Berquand, Léon, *Über das Stottern und dessen Heilung.* Vienna: Verlag des Wiener Magistrates, 1895.

Cazalet, Rev., *On the Right Management of the Voice in Speaking and Reading.* London: Bosworth & Harrison, 1860.

Clericus, P., *Die Pflege der Kinderstimme und der Stimmwechsel.* Berlin: Selbstverlag des Verfassers, 1906.

Coën, Rafael, *Das Stottern, Stammeln, Lispeln und die übrigen Sprechfehler sowie das Wesen, die Verhütung und Heilung dieser Uebel.* Vienna and Leipzig: A. Hartleben's Verlag, 1877.

Czech, Franz Hermann, *Versinnlichte Denk- und Sprachlehre mit Anwendung auf die Religions- und Sittenlehre und auf das Leben.* Vienna: Commission der Mechitaristen Congregations Buchhandlung, 1836.

Denhardt, Rudolf, *Das Stottern.* Leipzig: Verlag von Ernst Keil's Nachfolger, 1890.

Denhardt, Rudolf, *Ein offenes Wort in Sachen "Emil Denhardtsen" gegen Albert Gutzmann, Taubstummenlehrer und Dr. med. Hermann Gutzmann in Berlin*. Eisenach: Hofbuchdruckerei, s.d.

Denhardt, Rudolf, *Was ist Stottern und wie soll es behandelt werden?*, Leipzig: Verlag von Ernst Keil's Nachfolger, s.d.

Dobers, M., *Die Bekämpfung des Stotterübels in den Schulen. Erläuterungen zu den Wandtafeln*. Wittenberg: R. Herrofés Verlag, 1895.

Engel, Eduard, *Ueber den Stimmumfang sechsjähriger Kinder und den Schulgesang*. Berlin: Verlagsanstalt und Druckerei Actien-Gesellschaft, 1889.

Engel, Gustav, *Die Consonanten der deutschen Sprache. Erster, theoretischer Theil*. Berlin: Verlag von G. A. Challier & Co, 1874.

Ernst, Paul, Stimmbildung. *Sächsische Schulzeitung. Organ Des Sächsischen Lehrervereins Und Den Sächs. Pestalozzi-Vereins*, 21 (May 1909).

Ernst, Robert, *Kein Stottern mehr! Eine suggestive Selbsthilfe zur dauernden Beseitigung des Leidens*. Berlin: Anstaltsverlag, s.d.

Ernst, Robert, *Prospect der Heilanstalt für Sprachstörungen*, s.d.

Eyrel, Franz, *Physiologie der menschlichen Tonbildung nach den neuesten Forschungen gemeinfasslich dargestell. Ein praktisches Handbuch zur Ausbildung der Stimme und Sprache aller Menschen*. Leipzig: F. M. Brockhaus, 1860.

Fack, M., Die Behandlung stotternder Schulkinder. *Pädagogisches Magazin. Abhandlungen Vom Gebiete Der Pädagogik Und Ihrer Hilfswissenschaften*, 1904.

Frank, L., *Über Angstneurosen und das Stottern*. Zürich: Art. Institut Orell Füssli, s.d.

Freund, C. S., *Labyrinthtaubheit und Sprachtaubheit. Klinische Beiträge zur Kenntniss der sogenannten subcorticalen sensorischen Aphasie sowie des Sprachverständnisses der mit Hörresten begabten Taubstummen*. Wiesbaden: Verlag von J. F. Bergmann, 1895.

Gerber, Paul Henry, *Die menschliche Stimme und ihre Hygiene: Sieben volkstümliche Vorlesungen*. Leipzig: B. G. Teubner, 1911.

Gerdts, A. E., *Die naturgemässe Heilung des Stotterns*. Frankfurt a.M.: Selbstverlag des Verfassers, 1870.

Godtsring, Otto, *Praktische Anleitung zur Beseitigung des Stotterns und zur Kräftigung der Sprechstimme für Seminaristen und Lehrer*. Leipzig: Verlag der Dürr'schen Buchhandlung, 1906.

Gruenbaum, Ferdinand, *Erklärung des Stotterns dessen Heilung und Verhütung*. Leipzig: Verlag des Reichs-Medicinal-Anzeigers B. Konegen, 1897.

Grützener, P., Physiologie der Stimme und Sprache. In *Handbuch der Physiologie der Bewegungsapparate*. Leipzig: Verlag von F. C. W. Vogel, 1879.

Guttmann, Oskar, *Gymnastik der Stimme, gestützt auf physiologische Gesetze. Eine Anweisung zum Selbstunterricht in der Uebung und dem richtigen Gebrauche der Sprach- und Gesangsorgane*. Leipzig: J. J. Weber, 1861.

Gutzmann, Albert, Bemerkungen zu den öffentlichen Heilkursen für stotternde Schulkinder. *Medizinisch-Pädagogischen Monatsschrift Für Die Gesamte Sprachheilkunde*, 3: (3–4), 1893.

Gutzmann, Albert, *Das Stottern und seine gründliche Beseitigung durch ein methodisch geordnetes und praktisch erprobtes Verfahren*. Berlin: Verlag von Elwin Staude, 1890.

Gutzmann, Albert, *Übungsbuch für die Hand der Schüler im Anschluss an das Werk "Das Stottern."* Berlin: Verlag von Elwin Staude, 1892.

Gutzmann, Hermann, *Die Verhütung und Bekämpfung des Stotterns in der Schule. Ein Beitrag zur Schulhygiene*. Leipzig: Verlag von Georg Thieme, 1889.

Gutzmann, Hermann, *Stimmbildung und Stimmpflege. Gemeinverständliche Vorlesungen*. Wiesbaden: Verlag von J. F. Bergmann, 1906.

Gutzmann, Hermann, Stimme und Sprache ohne Kehlkopf. *Verhandlungen Des I. Internationale Laryngo-Rhinologen-Kongresses*, Vienna: Franz Deuticke, 1908.

Gutzmann, Hermann, Über Gewöhnung, und Gewohnheit, Übung und Fertigkeit, und ihre Beziehungen zu Störungen der Stimme und Sprache. In *Fortschritte der Psychologie und ihrer Anwendungen*. Leipzig: Verlag von B. G. Teubner, 1913.

Gutzmann, Hermann, Über Stottern als Herdsymptom. *Medizinisch-Pädagogischen Monatsschrift Für Die Gesamte Sprachheilkunde*, 20 (1910), 248–269.

Haase, C. A., *Das Stottern oder Darstellung und Beleuchtung der wichtigsten Ansichten über Wesen, Ursache und Heilung desselben nebst Abhandlung des Hieronymus Mercurialis "De Balbutie."* Berlin: In Commission bei August Hischwald, 1846.

Hagemann, Henriette, *Die untrügliche Heilung des Stotter- und Stammel-Uebels nach sechszehnjähriger Behandlung desselben*. Breslau: Verlag von Eduard Trewendt, 1845.

Hemmen, N., *Das Stottern. Statistik, Folgen, Abhülfe. Ein Beitrag zur Heilpädagogik*. Luxemburg: Druck der St. Paulus-Gesellschaft, 1890.

Hoffmann, Alfred, *Krankheiten der Sprache und Stimme. Was muss der Erzieher von ihrem Wesen und von ihrer Heilung wissen?* Meissen: Sächsische Schulbuchhandlung, 1910.

Hoffmann, P., *Lesebuch für den Stimmbildungsunterricht*. Halle: Druck und Verlag von Gebauer-Schwetschete, 1910.

Imhofer, Richard, *Die Ermüdung der Stimme (Phonasthenie)*. Würzburg: Verlag von Curt Kabitzsch, 1913.

Imhofer, Richard, *Die Krankheiten der Singstimme für Ärzte*. Basel: Verlag von Otto Enslin, 1904.

Jacobs, Theodor, *Dauernde Heilung für Stotternde*. Basel: Buchdruckerei zum Basler Berichthaus, 1903.

Kafemann, R., *Die Erkrankungen der Sprechstimme, ihre Ursachen und Behandlung nebst einer kurzen Hygiene für Lehrer, Geistliche, Advokaten und Offiziere*. Danzig: Verlag und Druck von A. W. Kafemann, 1899.

Kafemann, R., *Ueber die Beziehungen gewisser Nasen- und Rachenleiden zum Stottern auf Grund von Schuluntersuchungen*. Danzig: Verlag der Schulbuchhandlung von A. W. Kafemann, 1891.

Klencke, Hermann, *Die Heilung des Stotterns nach meiner rationellen, didactisch-medizinischen Methode*. Leipzig: Verlag von Ehr. Ernst Kollmann, 1860.

Kreuker, Fr., *Lehrbuch für Stotternde, eine Handreichung für Anstalten und Lehrer und zum Selbstgebrauch*. Rostock: Selbstverlag des Verfassers, s.d.

Krüger, Albracht, *Die menschliche Stimme, ihre Organe, Ausbildung, Pflege und Erhaltung*. Leipzig: Verlag von Kunst und Wissenschaft, 1898.

Kudelka, Josef, *Analyse der Laute der menschlichen Stimme vom physikalisch-physiologischen Standpunkte*. Linz: Druck und Verlag van J. Leichtinger's Erben, 1856.

Kussmaul, Adolf, *Versuch einer Pathologie der Sprache*, Leipzig: F.C.W. Vogel, 1885.

Lehwess, Joseph, *Radicale Heilung des Stotterns unter Anwendung der Respirations- und Sprach-Gymnastik*. Braunschweig: Druck und Verlag von Friedrich Vieweg und Sohn, 1868.

Lichtinger, *Ueber die Natur des Stotterns über die Wichtigkeit der medizinischen Behandlung desselben der chirurgischen gegenüber und über die physiologische Begründung der didaktischen Methode*. Berlin: A. W. Schade, 1844.

Liebing, Arno, *Gesundheitsgemässes und phonetisch richtiges Sprechen. Zur Aufklärung für Stimmleidende und Freunde einer rationellen Ausbildung der Stimme in Sprache und Gesang*. Dresden: Holze & Pahl, 1909.

Liebmann, Albert, *Stottern und Stammeln. Ihre Ursachen, Verhütung und Heilung*. Berlin: Verlag von Hugo Steinitz, 1895.

Liebmann, Albert, *Stotternde Kinder*. Berlin: Verlag von Reuther & Reichard, 1903.

Liebmann, Albert, *Vorlesungen über Sprachstörungen*. Berlin: Oscar Coblentz, 1898.

Liskovius, Karl Friedrich S., *Physiologie der menschlichen Stimme für Ärzte und Nichtärzte*. Leipzig: Verlag von Joh. Amb. Barth, 1846.

Loessin, Paul, *Stottern. Behandlung zur selbsteigenen radikalen Beseitigung des Stotterns und Erlangung einer normalen, fliessenden Redeweise*. Berlin: Otto Dreyer, s.d.

Makowsky, Lew, *Beitrag zur Lehre von der Localisation des Sprachcentrums*. Jena: Julius Hossfeld, 1883.

Malebouche, Fortuné, *Précis sur les causes du bégaiement et sur les moyens de le guérir*. Paris: Fortin, Masson etc., 1841.

Mehnert, Max, *Über Sprachstörungen mit besonderer Berücksichtigung des Stammelns und Stotterns bei Schulkindern*. Dresden: Gebr. Adolph & Co., 1904.

Morill, Bianka, *Stimmerziehende Lautbildungslehre nach einem Lautbildungsgesetz*. Berlin-Gross Lichterfelde: Chr. Friedrich Vieweg, 1907.

Mutke, Robert, *Die Behandlung stammelnder und stotternder Schüler*. Breslau: Verlag von Franz Overlich, s.d.

Müller, Johannes, *Über die Compensation der physischen Kräfte am menschlichen Stimmorgan. Mit Bemerkungen über die Stimme der Säugethiere, Vögel und Amphibien*. Berlin: A. Hirschwald, 1839.

Nadolezny, Max *Lautbildung und Sprachstörungen mit Berücksichtigung der Stimmhygiene*. Munich: Verlag der Aerztlichen Rundschau Otto Gmelin, 1912.

Neumann, Albert, *Heilübungsbuch für Stotternde und Stammelnde etc. Für Lehrer, Eltern und Erzieher, sowie zum Gebrauche für Erwachsene*. Graz: Albert Neumann's Verlag, 1893.

Nikolaisen, Nicolai, *Praktische Anleitung zur Beseitigung des Stotterns für Lehrer und Eltern*. Flensburg: Verlag von Aug. Westphalen, 1887.

Otto, F. *Das Geheimnis Stotternde und Stammelnde zu heilen für Eltern, Erzieher, Lehrer und Ärzte*. Halle: Eduard Anton, 1832.

Pestalozzi, Heinrich, *Zur Vermeidung von Irrwegen bei der Ausbildung der menschlichen Stimme durch Individuelle Stimmbildung*. Berlin: Albert Stahl, 1910.

Pick, A., Zur Lehre vom Einfluss des Sprechens auf das Denken. *Zeitschrift Psychologische und Physiologische Sinnesorg*, 44 (1907), 291–295.

Piper, Hermann, *Der kleine Sprachmeister: ein Lehr- und Bilderbuch*. Berlin: Verlag von Karl Siegismund, 1897.

Päzolt, Theodor, *Die seelischen Hemmungserscheinungen des Stotterns. Eine pädagogisch-psychologische Studie*. Leipzig: Verlag des Verfassers, 1903.

Radau, J. C., *Religions-Büchlein nach Luthers Catechismus für taubstumme Kinder*. Augerburg: Druck und Verlag van J. J. Priddat, 1863.

Rosenthal, Moritz, *Beitrag zur Kenntnis und Heilung des Stotterübels*. Vienna: Sallmayer & Comp., 1861.

Rossbach, Michael Joseph, *Physiologie und Pathologie der menschlichen Stimme auf Grundlage der neuesten akustischen Leistungen*. Würzburg: A. Stuber's Buchhandlung, 1869.

Rutz, Ottmar, *Neue Entdeckungen von der menschlichen Stimme*. Munich: C. H. Beck'sche Verlagsbuchhandlung, 1908.

Sandow, L. *Mechanik des Stotterns. Gründliche Selbstheilung ohne Atem-, Artikulations'-, Stimmbildung's- und Sprechübungen*. Nordhausen: E. Edler, 1898.

Scharr, Julius, *Der Sprechorganismus, die wichtigsten Sprechfehler (Stottern und Stammeln) und deren Heilung durch die Schule. Ein Hilfsbuch in der Anthropologie und Lautlehre für Lehrer und Seminaristen*. Vienna and Leipzig: Verlag von A. Pichler's Witwe und Sohn, 1897.

Scharr, Julius, *Die Behandlung Stotternder. Eine Anleitung für Lehrer und Leiter von Stotterkursen.* Hanover: Norddeutsche Verlagsanstalt, 1893.

Scheidemantel, Karl, *Stimmbildung* (2nd ed.). Leipzig: Breitkopf & Härtel, 1908.

Schilling, Kurt, Zur Stimmbildungsfrage. *Sächsische Schulzeitung. Organ Des Sächsischen Lehrervereins Und Den Sächs. Pestalozzi-Vereins*, 16 (1910), 253–255.

Schneider, Matthias, *Das Denken und das Sprechen des Taubstummen. Eine Untersuchung über die wahren Grundlagen des Taubstummen-Unterrichtes.* Osterwieck: A. W. Zickfeldts Verlag, 1908.

Schneider, Peter Joseph, *Fragmente aus dem Tagebuch betreffende meine Methode, eine Schrift für Jedermann.* Bonn and Cologne: Renard und Dubnen, 1835.

Schrank, Joseph, *Das Stotterübel eine corticale Erkrankung des Grosshirnes.* Munich: Verlag von Jos. Anton Finsterlin, 1877.

Schulthess, Rudolf, *Das Stammeln und Stottern. Über die Nature, Ursachen und Heilung dieser Fehler der Sprache.* Zürich: bei Friedrich Schulthess, 1830.

Schwidop, O., *Sprache, Stimme und Stimmbildung*, Karlsruhe: J.J. Reiff, 1898.

Sprechen, Singen, Musik. Grund- und Zeitfragen aus der Stimmkunde, der Musiktheorie und der musikalischen Jugend- und Volkserziehung, Leipzig: Bürrsche Buchhandlung, 1914.

Ssikorski, J. A., *Über das Stottern.* Berlin: Verlag von August Hirschwald, 1891.

Stötzner, Mein erster Heilversuch, *Medizinisch-Pädagogischen Monatsschrift für die Gesamte Sprachheilkunde*, XVII:11 (1907).

Urteile von Ärzten, Behörden, Lehrern, Geistlichen, Offizieren, Künstlern und Anderen über die Stimmbildungslehre Professor Eduard Engel's. Burg b.M.: August Hopfer, 1911.

Vogler, S., and Pflips, H., *Stottern heilbar durch eine neue, eigene Methode. Anleitung zur Behandlung durch Eltern und Lehrer, sowie zur Selbstbehandlung.* Cologne: Verlag von Friedr. Kratz & cie, s.d.

Von Meyer, Georg Hermann, *Stimm- und Sprachbildung.* Berlin: Lüderitsche Verlagsbuchhandlung Carl Habel, 1871.

Warnecke, [Ludwig] *Prospekt der internationalen Sprachheil-Anstalt (Abteilung für Patentverwertung) Hannover*, s.d.

Weiss, G. Gottfried, *Allgemeine Stimmbildungslehre für Gesang und Rede mit anatomisch physiologischer Begründung.* Braunschweig: Friedrich Vieweg und Sohn, 1868.

Widmann, Benedict, *Gehör- und Stimmbildung. Eine auf physiologische, psychologische und pädagogische Untersuchungen und Beobachtungen gegründete Anleitung zur Pflege des Gehörs und der Stimme* (2nd ed.). Leipzig: Carl Merzeburger, 1899.

Wolff, Gustav, *Klinische und kritische Beiträge zur Lehre von den Sprachstörungen.* Leipzig: Verlag von Veit & Comp., 1904.

Wolff, I. *Schnelle Heilung des Stotterns.* Berlin: Verlag von Theobald Grieben, 1878.

Leipzig: Hochschule für Musik und Theater Felix Mendelssohn

Armin, George, *Das Stauprinzip oder die Lehre von dem Dualismus der menschlichen Stimme dargelegt für Sänger, Schauspieler und Rezitatoren.* Leipzig: C. F. W. Siegel's Musikalienhandlung, 1909.

Baumann, Friedrich, *Die Ausbildung der Kehle zum Instrument : eine kurze und gründliche Anweisung, richtig zu singen; zum Gebrauche für Gesangslehrer, Sänger, Seminare & Schullehrer, so wie auch zum Selbstunterrichte.* Leipzig: Verlags-Contor, 1859.

Baumann, Friedrich, *Das Unentbehrlichste aus der Gesangskunst eine kurze und gründliche Anweisung die Stimme gut erhalten und richtig gebrauchen zu lernen ; nach rationellen Grundsätzen.* Munich: Selbstverlag des Verfassers, 1864.

Böhme-Köhler, A., *Lautbildung beim Singen und Sprechen. Ein Leitfaden zum Unterricht in Schulen und für Privatgebrauch.* Leipzig: Friedrich Brandstetter, 1905.

Bottermund, W., *Die Gesundheitspflege der Stimme, des Gesanges und der Sprache.* Leipzig: C. F. W. Siegel's Musikalienhandlung, n.d.

Bruns, Paul, *Das Problem der Kontra-altstimme.* Berlin: Gross Lichterfelde, 1906.

Carlberg, Gotthold, *Die Kunst, Sänger zu werden Eine auf Erfahrung begründete Anleitung, besonders für Dilettanten, sich in der Kunst des Gesanges, in Bezug auf Stimme und Vortrag durch Selbstunterricht zu vervollkommen.* A. Hartleben's Verlag, 1870.

Cludius, M. R., *Grundris der körperlichen Beredsamkeit für Liebhaber der schönen Künste, Redner und Schauspieler.* Hamburg: Carl Ernst Bohn, 1792.

D'Aubigny von Engelbrunner, Nina, *Briefe an Natalie über den Gesang : als Beförderung der häuslichen Glückseligkeit und des geselligen Vergnügens ; ein Handbuch für Freunde des Gesanges, die sich selbst, oder für Mütter und Erzieherinnen, die ihre Zöglinge für diese Kunst bilden möchten.* Leipzig: Voss, 1803.

Ephraim, Alfred, *Die Hygiene des Gesanges. Für Sänger und Gesanglehrer dargestellt.* Leipzig: Breitkopf & Härtel, 1899.

Escudier, Marie, *Études biographiques sur les chanteurs contemporains : précédées d'une esquisse sur l'art du chant.* Paris: Tessier, 1840.

Flatau, Theodor S., *Die funktionelle Stimmschwäche (Phonastenie) der Sänger, Sprecher und Kommandorufer.* Charlottenburg: Georg Bürkners Verlag, 1906.

Hacke, Heinrich, *Lerne singen!* Berlin: Lorelei Verlag, 1905.

Haeser, Heinrich, *Die menschliche Stimme, ihre Organe, ihre Ausbildung, ihre Pflege und Erhaltung für Sänger, Lehrer und Freunde des Gesanges.* Berlin: Verlag von August Hirschwald, 1839.

Hagemann, Carl, *Der Mime. Die Kunst des Schauspielers und Opernsängers.* Berlin: Schuster & Loeffler, 1918.

Härtinger, Martin Franz, *Die menschliche Stimme.* Munich: Johan Deschler, 1840.

Hoerger, Joseph, *Kurze und faßliche Sing-Anleitung.* Augsburg: Boehm, 1822.

Hohmann, Christian Heinrich, *Praktischer Lehrgang für den Gesang-Unterricht.* Nördling: C. H. Beck, 1852.

Jakob, Friedrich A. L., *Fassliche Anweisung zum Gesang-Unterricht in Volksschulen: Nach naturgemäßen Grundsätzen und das Singen nach Noten und Ziffern.* Breslau: Grüson, 1828.

Koehler, Louis, *Gesangs-Führer:(Aus Schafhäutls' Nachlass) Ein Auszug empfehlenswerter Werke aus der gesammten Literatur für Solo- und Chorgesang.* Leipzig: Schuberth's Buchhandlung, 1863.

Könen, Peter, *Unterricht in dem Gregorianischen Chorgesang : nebst Anweisung die deutschen Kirchenlieder nach der Musik zu singen, und in Choral-Noten zu übersetzen.* Cologne, 1810.

Kornmüller, Utto, *Der katholische Kirchenchor oder die vorzüglichsten Pflichten der Chorregenten, Organisten und Sänger.* Landshutt: Thomann, 1868.

Kösporer, Johann Nepomuk, *Leitfaden für den Gesang-Unterricht.* Freysing: Datterer, 1861.

Kürzinger, Ignaz Franz Xaver, *Getreuer Unterricht zum Singen mit Manieren und die Violin zu spielen.* Augsburg: Lotter, 1821.

Martienssen, Franziska, *Das bewußte Singen : Grundlegung des Gesangstudiums.* Leipzig: Kahnt, s.d.

Marx, Adolf Bernhard, *Die Kunst des Gesanges : theoretisch-praktisch*, Berlin: Schlesinger, 1826.

Merling, Julius, *Theoretisch-praktischer Gesangs-Cursus : Zum Gebrauch in höheren und mittleren Schulen, in Seminarien, beim Privatunterrichte für Lehrende und Lernende nach vier Stufen bearbeitet*. Magdeburg: Heinrichshofen, 1863.

Minoja, Anastas, *Anleitung zur vollständigen Ausbildung im Gesange Nebst Lehre das Organ zu kräftigen, dauernd zu verschönern und den höchsten Wohlklang der Stimme zu erzielen*. Leipzig: Verlag von Moritz Ruhl, 1857.

Mosewius, Johann Theodor, *Die Breslauische Sing-Akademie in den ersten fünf und zwanzig Jahren ihres Bestehens*. Breslau: Schuhmann in Komm, 1850.

Nehrlig, C. G., *Die Gesangkunst, physiologisch, psychologisch, ästhetisch und pädagogisch dargestellt. Anleitung zur vollendeten Ausbildung im Gesange, sowie zur Behandlung und Erhaltung des Stimmorgans und zur Wiederbelebung eines verlosen geglaubten Stimme*. Leipzig: B. G. Teubner, 1853.

Noë, Oskar, and Moser, Hans Joachim, *Technik der deutschen Gesangskunst*. Leipzig: G. J. Göschen'sche Verlagshandlung, 1912.

Ortloph, W., *Theoretisch-praktische Gesang-Schule*. Munich: D. Carl Wolff, 1842.

Pelzer, Eduard, *Gesang. Handbuch zur wissenschaftlichen Fortbildung der Lehre und zur Vorbereitung auf die zweite Lehrer- Mittelschul- und Rektorprüfung*. Düsseldorf: L. Schwan, 1909.

Pruckner, Caroline, *Theorie und Praxis der Gesangskunst : Handbuch für angehende Sänger und Sängerinnen*. Vienna: Czermak, 1872.

Renner, Friedrich, *Gesang-schule*. Leipzig: Constantin Wild's Verlag, 1892.

Reusch, Fritz, *Der kleine Hey. Die Kunst des Sprechens. Nach dem Urtext von Julius Hey*. Mainz: B. Schott's Söhne, 1956.

Schebest, Agnese, *Aus dem Leben einer Künstlerin*. Stuttgart: Ebner & Seubert, 1857.

Scheidemantel, Karl, *Gesangsbildung*. Leipzig: Breitkopf & Härtel, s.d.

Schilling, Gustav, *Musikalische Dynamik oder die Lehre vom Vortrage in der Musik : ein Lehr-, Hand- und Hülfsbuch für alle, die auf irgend eine Weise praktische Musik treiben, Künstler oder Dilettanten, Sänger oder Instrumentalisten, Lehrer und Schüler*. Kassel: Krieg, 1843.

Schilling, Kurt, *Schulpraktische Richtlinien für Singen, Sprechen un Rechtschreiben*. Meissen: Sächsische Schulbuchhandlung, 1915.

Schipke, Max, *Der deutsche Schulgesang von Johann Adam Hiller bis zu den Falkschen Allgemeinen Bestimmungen (1775–1875)*. Berlin: Union Deutsche Verlagsgesellschaft, 1913.

Schladebach, Julius, *Die Bildung der menschlichen Stimme zum Gesang Mit Holzschnitten*. Sondershausen: Verlag von G. Reuse, 1860.

Schmidt, Heinrich, *Gesang und Oper: kritisch-didaktische Abhandlungen in zwanglosen Heften*. Magdeburg: Heinrichshofen, 1861.

Schubert, Johann Friedrich, *Neue Singe-Schule oder gründliche und vollständige Anweisung zur Singkunst: in drey Abtheilungen mit hinlänglichen Uebungsstücken*. Leipzig: Breitkopf & Härtel, 1790.

Schwarz, Wilhelm, *Die Musik als Gefühlssprache im Verhältniss zur Gesang- und Stimmbildung*. Leipzig: C. F. Kahnt, 1860.

Smita, Josef, *Singen und Sprechen: ein Beitrag zum Gesangsunterrichte an Gymnasien mit Rücksicht auf den*. Teschen: Hofbuchdruckerei Karl Prochaika, 1884.

Töpfer, J. G., *Die Scheib'lersche Stimmmethode*. Erfurt: Wilhelm Körner, 1842.

Van Zanten, Cornelie, *Leitfaden zum Kunstgesang*. Leipzig: Breitkopf & Härtel, 1903.

Wagner, Richard, *Über Schauspieler und Sänger*. Leipzig: Fritzsch, 1872.

Wicke, Richard (ed.), *Sprechen, Singen, Musik. Grund-und Zeitfragen aus der Stimmkunde, der Musiktheorie und der musikalischen Jugend- und Volkserziehung.* Leipzig: Dürssche Buchhandlung, 1914.

Wieck, Friedrich, *Clavier und Gesang: Didaktisches und Polemisches.* Leipzig: Whistling, 1853.

Zopff, Hermann, *Grundzüge einer Theorie der Oper: ein theoretisch-praktisches Handbuch für Künstler und Kunstfreunde, Dichter und Componisten, Sänger, Capellmeister, Regisseure und Directoren.* Leipzig: Arnoldi, 1868.

London: Wellcome Collection

ARCHIVES RELATED TO MORRELL MACKENZIE

Czermak, James, *Letter to Mackenzie, Leipzig Jan. 5th 1870.*

Frederick, V. Empress, *Letter to Morrell Mackenzie, 8 January 1889.*

Garcia, Manuel, *Letter to Mackenzie, 19 april 1887.*

Garcia, Manuel, *Letter to Mackenzie, le 3 août 1887.*

Garcia, Manuel, *Note to Mackenzie, undated.*

Mackenzie, Morrell, *Photographs of Throat*, 1890.

O'Connor, E. I. J., *Letter to Mackenzie, Jan 17th, 1889.*

Priestly, William O., *Letter to Mackenzie, January 27, 1875.*

Prussia, Frederick of, *Emperor Frederick's Notes*, 1888.

Reid, James, *Letter to Morrell Mackenzie, 29 November 1887.*

Reid, James, *Letter to Mackenzie, 23 January 1888.*

Sims, Marion, *Letter to Mackenzie, Nov. 29, 1882.*

PATENTS AND COMMERCIAL SOURCES

Mitford, Bertram, *Communicating with Deaf and Dumb Persons* (Patent No. 1075). Great Seal Patent Office, 1868.

Mocatta, Percy G., and Allen, A., *The Lost Voice (A Song of the "Ammoniaphone").* The Medical Battery Company, 1880.

Rehse, John, *System for the Cure of Stammering* (Patent No. 2190), 1870.

PUBLISHED SOURCES

"On Stammering," *Medico-Chirurgical Review, 25: 50* (1836), pp. 319–322

Georg Hermann von Meyer, *The Organs of Speech and Their Application in the Formation of Articulate Sounds*, London: Kegan Paul, Trench & Co, 1883.

Abbots, W., *Stammering, Stuttering and Other Speech Affections: Their Causes and Cure.* London: The Savoy Press, 1894.

Aikin, W. A., *The Voice. An Introduction to Practical Phonology.* London: Longmans Green, 1910.

Alexander, F. Matthias, *Introduction to a New Method of Respiratory Vocal Re-Education.* London: Baillier, Tindall and Cox, 1906.

Alison, S. Scott, *Morbid Conditions of the Throat in Their Relation to Pulmonary Consumption.* London: H. K. Lewis, 1869.

Appelt, Alfred, *The Stammerer's Guide.* Manchester: Harry Glynn & Co., 1905.

Appelt, Alfred, *The Real Cause of Stammering and Its Permanent Cure. A Treatise on Psycho-Analytical Lines.* London: Methuen & Co., 1911.

Ball, James B., *Intubation of the Larynx.* London: H. K. Lewis, 1891.

Barnes, Harry A. *The Tonsils. Faucial, Lingual and Pharyngeal. With Some Account of the Posterior and Lateral Pharyngeal Nodules. Illustrated.* London: Henry Kimpton, 1914.

Bartlett, T. *Stammering Practically Considered, with the Treatment in Detail.* London: Sherwood, Gilbert and Piper 1839.

Barwell, Harold, *Diseases of the Larynx*, London: Oxford University Press, 1907.

Bataille, Charles, *Nouvelles recherches sur la phonation. Mémoire présenté et lu à l'académie des sciences le 15 avril 1861.* Paris: Victor Masson et fils, 1861.

Beasley, Benjamin, *Stammering: Its Treatment.* London: Waterlow and Sons Limited, 1888.

Beasley, Benjamin, *Reminiscences of a Stammerer.* London: The Roxburghe Press, 1902.

Becquerel, A. *Traité du bégaiement et des moyens de le guérir.* Paris: Fortin, Masson etc., 1843.

Behnke, Emil, and Lennox Browne, *The Child's Voice: Its treatment with Regard to After Development.* Chicago: A. N. Marquis & Company, 1885.

Bell, Alexander Melville, *Visible Speech. The Science of Universal Alphabetics; or Self-Interpreting Physiological Letters, for the Writing of All Languages in One Alphabet. Illustrated by Tables, Diagrams, and Examples.* London: Simpkin, Marshall & Co., 1867.

Bell, Alexander Graham, *Lectures upon the Mechanism of Speech. Reprinted from the Proceedings of the First Summer Meeting of the American Association to Promote the Teaching of Speech to the Deaf.* New York: Funk & Wagnalls Company, 1906.

Bell, Alexander Melville, *The Science of Speech.* Washington, D.C.: Volta Bureau, 1897.

Bennett, Edmund F. T., *A Self Cure of Stammering, Stuttering and Indistinct Speech.* London: Simpkin, Marshall, Hamilton, Kent & Co., 1905.

Bishop, John, *On Articulate Sound; and on the Causes and Cure of Impediments of Speech.* London: Samuel Highley, 1860.

Bristowe, John Syer, *The Physiological & Pathological Relations of the Voice and Speech.* London: David Bogue, 1880.

Broca, Paul, "Du siège de la faculté du langage articulé dans l'hémisphère gauche du cerveau." *Bulletins de La Société d'anthropologie*, 6 (1865), 1–18.

Browne, Lennox, *Science and Singing.* London: Chappell & Co., 1880.

Browne, Lennox, *Voice Use and Stimulants.* London: Sampson Low, Marston, Searle, and Rivington, 1885.

Browne, Lennox, and Emil Behnke, *Voice, Song and Speech: A Practical Guide for Singers and Speakers, from the Combined View of Vocal Surgeon and Voice Trainer. With Numerous Illustrations by Wood-Engraving and Photography* (3rd ed.). London: Sampson Low, Marston, Searle, and Rivington, 1884.

Browne, Lennox, *The Throat and Its Diseases. With One Hundred Typical Illustrations in Color, and Fifty Wood Engravings, Designed and Executed by the Author.* London: Baillier, Tindall and Cox, 1878.

Browne, Lennox, *Koch's Remedy in Relation Specially to Throat Consumption. Illustrated by Thirty-One Cases and by Fifty Original Engravings and Diagrams.* London: Baillier, Tindall and Cox, 1891.

Brücke, Ernst, *Grundzüge der Physiologie und Systematik der Sprachlaute für Linguisten und Taubstummenlehrer.* Vienna: Carl Gerold's Sohn, 1856.

Butlin, Henry T., *On Malignant Disease (Sarcoma and Carcinoma) of the Larynx.* London: J. & A. Churchill, 1883.

Cassanello de Montevideo, Eugenio, *Cases of Cured Tracheotomy* (medical photography), 1883.

Collins, Joseph, *The Genesis and Dissolution of the Faculty of Speech. A Clinical and Psychological Study of Aphasia.* New York: Macmillan Company, 1898.

Colombat de l'Isere, Marc, *L'orthophonie ou physiologie et thérapeutique du bégaiement et de tous les vices de la prononciation.* (2nd ed.). Paris: Mansut fils, 1833.

Colombat de l'Isère, Marc, *Traité des maladies et de l'hygiène des organes de la voix*, Manut, 1838

Colombat de l'Isere, Marc, *A Treatise upon the Diseases and Hygiene of the Organ of the Voice* (J. F. W. Lane, M.D., trans.; 2nd ed.). Boston: Redding & Company, 1857.Court de Gébelin, Antoine, *Histoire naturelle de la parole ou grammaire universelle à l'usage des jeunes gens.* (2nd ed.). Paris: Plancher, Eymery, Delaunay, 1816.

Cull, Richard, *Stammering, Considered with Reference to Its Cure by the Application of Those Laws Which Regulate Utterance. In a Letter Addressed to George Birkbeck*, London: Henry Renshaw, 1835.

De Gembloux, Pierquin, *Idiomologie des animaux ou recherches historiques, anatomiques, physiologiques, philologiques, et glossologiques sur le langage des bêtes.* Paris: A la tour de Babel, 1844.

de la Chapelle, M., *Le ventriloque ou l'engastrimythe.* London: Chez de l'Etanville, 1772.

Downie, James Walker, *Clinical Manual for the Study of Diseases of the Throat.* Glasgow: James Maclehose and Sons, 1894.

Ernst, Friedrich, *Die Krankheiten der Nase und des Halses ihre Beziehungen zum Gesammtorganismus und ihre Bedeutung für die Singstimme.* Berlin: Verlag von Adolf W. Köllner, 1899.

Farrar, James, *The Human Voice and Connected Parts. A Practical Book for Orator, Clergymen, Vocalists and Others.* London: Marshall Japp & Company, 1881.

Faulkner, Richard B., *The Tonsils and the Voice in Science, Surgery, Speech and Song. A Comprehensive Monograph on the Structure, Utility, Derangements and Treatment of the Tonsils, and of Their Relationship to Perfect Tone Production. A Research Study with Original Contributions.* Pittsburgh: Presbyterian Book Store, 1913.

Fauvel, Charles, *Traité pratique des maladies du larynx précédé d'un traité complet de laryngoscopie.* Paris: Adrien Delahaye, Libraire-éditeur, 1876.

Fournié, Édouard, *Physiologie de la voix et de la parole. Avec figures dans le texte.* Paris: Adrien Delahaye, Libraire-éditeur, 1866.

Gardiner, William, *The Music of Nature or an Attempt to Prove That What Is Passionate and Pleasing in the Art of Singing, Speaking, and Performing upon Musical Instruments Is Derived from the Sounds of the Animated World. With Curious and Interesting Illustrations.* London: Longman, Rees, Orme, Brown, Green and Longman, 1832.

Garner, Richard Lynch, *The Speech of Monkeys.* London: William Heinemann, 1892.

Garrod, Archibald E., *An Introduction to the Use of the Laryngoscope.* London: Longmans, Green and Co., 1886.

Gibb, George Duncan, *On Diseases of the Throat, Epiglottis and Windpime; Including Diphteria, Nervous Sore-Throat, Displacements of the Cartilages, Weakness of the Voice and Chest: Their Symptoms, Progress and Treatment.* London: John Churchill, New Burlington Street, 1860.

Gibb, George Duncan, *The Laryngoscope: Illustrations of its Practical Application, and Description of Its Mechanism.* London: John Churchill, New Burlington Street, 1863.

Gibb, George Duncan, *On Diseases of the Throat and Windpipe, as Reflected by the Laryngoscope. A Complete Manual upon their Diagnosis and Treatment. Embellished with 116 Engravings.* (2nd ed.). London: John Churchill, New Burlington Street, 1864.

Gibb, George Duncan, *The Laryngoscope in Diseases of the Throat with a Chapter on Rhinoscopy. A Manual for the Student and Practitioner.* (3rd ed.). London: John Churchill, New Burlington Street, 1868.

Gottstein, J., *Diseases of the Larynx* (trans.). Edinburgh: W. & A. K. Johnston, 1885.

Gottstein, J., *Die Krankheiten des Kehlkopfes. Mit Einschluss der Laryngoskopie und der localtherapeutischen Technik für praktische Aerzte und Studirende.* Leipzig: Franz Deuticke, 1890.

Gouguenheim, Achille, and Tissier, Paul, *Phtisie laryngée.* Paris: Masson, 1889.

Grünwald, L., *Atlas and Abstract of the Diseases of the Larynx.* London: Rebman Publishing Co., 1898.

Grünwald, L., *Grundriss der Kehlkopfkrankheiten und Atlas der Laryngoskopie. Mit 112 farbigen Abbildungen auf 47 Tafeln und 26 schwarzen Abbildungen im Text.* (2nd ed.). Munich: J. F. Lehmann's Verlag, 1907.

Gutzmann, Hermann, Behandlung der Sprachstörungen. In: F. Penzholdt & R. Stintzing (eds.), *Handbuch der Therapie innerer Krankheiten,* Jena: Gustav Fischer, 1898, 503–523.

The Harangues, or Speeches, of Several Celebrated Quack-Doctors in Town and Country, London: Printed for J. Thomson in the Strand, 1762.

Hartley, A. M., *Observations on the Different Kinds of Impediments of Speech, Particularly Spasmodic Stammering: With Instructions How to Remove Them.* London: Simpkin, Marshall & Co., 1833.

Head, Henry, Special Issue: This Number Is Devoted to Hughling Jackson's Views of Apahsia and Other Allied Affections of Speech. *Brain. A Journal of Neurology, 38* (I & II). (1915).

Helmore, Frederick, *Speakers, Singers and Stammerers.* London: Joseph Masters and Co., 1874.

Heryng, Theodor, *Die Heilbarkeit der Larynxphthise und ihre chirurgische Behandlung. Eine anatomo-pathologische und klinische Studie.* Stuttgart: Verlag von Ferdinand Enke, 1887.

Hollander, Bernard, *A Brain Seen from the Underside; with Attention to the Area Associated by Gall with Verbal Memory.* London: Grant Richards, 1901.

Holmes, Gordon, *A Treatise on Vocal Physiology and Hygiene with Especial Reference to the Cultivation and Preservation of the Voice.* London: J. & A. Churchill, 1879.

Hulbert, H. H., *Voice Training in Speech and Song. An Account of the Structure and Use of the Vocal Organs and the Means of Securing Distinct Articulation.* (3rd ed.). London: W. B. Clive. University Tutorial Press, 1910.

Hunt, James, *A Treatise on the Cure of Stammering, with a General Account of the Various Systems for the Cure of Impediments in Speech and a Notice of the Life of the Late Thomas Hunt.* (3rd ed.). London: Longman, Green, Longman, and Roberts, 1857.

Hunt, James, *Stammering and Stuttering, Their Nature and Treatment.* London: Longman, Green, Longman, and Roberts, 1861.

Johnson, George, *The Laryngoscope. Directions for Its Use, and Practical Illustrations of Its Value in the Diagnosis and Treatment of Diseases of the Throat and Nose. Two Lectures Delivered at the Royal College of Physicians.* London: Robert Hardwicke, 1864.

Ketley, William J., *Stammering: The Beasley Treatment.* Birmingham: Hudson and Son, 1900.

Kirstein, Alfred, *Autoscopy of the Larynx and the Trachea. Direct Examination Without Mirror.* Philadelphia: The F. A. Davis Co., 1897.

Klindworth, L. *Stuttering and How to Cure It: A Practical and Systematical Handbook for Self-Instruction; Containing an Approved Method of Respiration and the Gymnastic of Voice and Speech.* Glasgow: F. Bauermeister, 1896.

Kofler, Leo, *The Art of Breathing as the Bases of Tone-Production. A Book Indispensable to Singers, Elocutionists, Educators, Barristers, Clergymen, and to All Others Desirous of Having a Pleasant Voice and Good Health.* London: J. Curwen & Sons, 1901.

Kussmaul, Adolf, *Störungen der Sprache. Versuch einer Pathologie der Sprache.* Leipzig: Verlag von F. C. W. Vogel, 1877.

Lake, Richard, *Laryngeal Phthisis or Consumption of the Throat*. London: Rebman, Limited, 1901.

Lane, Sir William Arbuthnot, *Cleft Palate and Hare Lip*. (3rd ed.). London: Adlard and Son, 1916.

Lee, Edwin, *On Stammering and Squinting, and on the Methods for Their Removal*. London: John Churchill, 1841.

Mackenzie, Morrell, *The Use of the Laryngoscope in Diseases of the Throat. With an Appendix on Rhinoscopy*. London: Robert Hardwicke, 1865.

Mackenzie, Morrell, *Du laryngoscope et de son emploi dans les maladies de la gorge avec un appendice sur la rhinoscopie* (trans.). Paris: J. B. Baillière et fils, 1867.

Mackenzie, Morrell, *Essay on Growths in the Larynx: With Reports, and an Analysis of One Hundred Consecutive Cases Treated by the Author and a Tabular Statement of All Published Cases Treated by Other Practictioners Since the Invention of the Laryngoscope*. London: J. & A. Churchill, 1871.

Mackenzie, Morrell, *On the Results of Thyrotomy for the Removal of Growths from the Larynx. Being a Reply to Mr. Durham's Paper in the Fifty-Fifth Volume of the Transactions of the Royal Medical and Chirurgical Society*. London: John Churchill, New Burlington street, 1873.

Mackenzie, Morrell, *The Hygiene of the Vocal Organs. A Practical Handbook for Singers and Speakers*. London: Macmillan and Co., 1886.

Mackness, James, *Dysphonia Clericorum or Clergyman's Sore-Throat: Its Pathology, Treatment, and Prevention*. London: Longman, Brown, Green and Longmans, 1848.

Macnamara, Nottidge C., *Human Speech. A Study in the Purposive Action of Living Matter*. London: Kegan Paul, Trench & Co., 1908.

Marage, René, *Petit manuel de physiologie de la voix à l'usage des chanteurs et des orateurs*. Paris: Chez l'auteur, 1911.

Marcet, William, *Clinical Notes on Diseases of the Larynx Investigated and Treated with the Assistance of the Laryngoscope*. London: Longmans, Green and Co., 1869.

Marchesi, Mathilde, *Marchesi and Music: Passages from the Life of a Famous Singing Teacher*, New York: Harper & Brother Publishers, 1897

Mason, Francis, *On Harelip and Cleft Palate*. London: J. & A. Churchill, 1877.

McCormac, Henry, *A Treatise on the Cause and Cure of Hesitation of Speech or Stammering*. London: Longman, Rees, Orme, Browne, and Green, 1828.

Merkel, Carl Ludwig, *Anatomie und Physiologie des menschlichens Stimm-und Sprachorgans (Anthropophonik). Nach eigenen Beobachtungen und Versuchen wissenschaftlich begründet und für studirende und ausübende Ärzte, Physiologen, Akustiker, Sänger, Gesanglehrer, Tonsetzer, öffent*. Leipzig: Verlag von Ambrosius Abel, 1857.

Mills, Wesley, *Voice Production in Singing and Speaking, Based on Scientific Principles*. London: J. Curwen & Sons, 1906.

Molinié, J., *Les tumeurs malignes du larynx. Pathologie - thérapeutique. Avec 65 figures dans le texte*. Paris: A. Maloine, 1907.

Monro, Henry, *On Stammering and Its Treatment*. London: John Churchill, 1850.

Monroe, Lewis B., *Manual of Physical and Vocal Training for the Use of Schools and for Private Instruction*. Philadelphia: Cowperthwait & Co., 1869.

Mott, Frederick W., *The Brain and the Voice in Speech and Song*. New York: Harper & Brothers, 1910.

Myer, Edmund J., *Vocal Reinforcement. A Practical Study*. Boston: American Publishing, 1891.

Müller, Johannes, *The Physiology of the Senses, Voice, and Muscular Motion, with the Mental Faculties*. London: Taylor, Walton, and Maberly, 1848.

Norris Wolfenden, R. *Studies in Pathological Anatomy, Especially in Relation to Laryngeal Neoplasms*, London: J. & A. Churchill, 1888.

Norton, Arthur Trehem, *Affections of the Throat and Larynx. The Classification, Description and Statistics of 150 Cases Occurring in the Throat Department of St. Mary's Hospital*. London: Robert Hardwicke, 1871.

Norton, Arthur Trehem, *Affections of the Throat and Larynx. Their Classification, Pathology, Diagnosis and Treatment*. (2nd ed.). London: Baillier, Tindall and Cox, 1875.

Parker, Robert William, *Tracheotomy in Laryngeal Diphtheria (Membranous Croup) with Especial Reference to After-Treatment to Which Are Added a Few General Remarks on Diphtheria and Its Earlier Treatment*. London: David Bogue, 1880.

Parker, Robert William, *Tracheotomy in Laryngeal Diphtheria. Aftertreatment and Complications*. (2nd ed.). London: H. K. Lewis, 1885.

Physician, Anonymous, *A Treatise on the Nature and Causes of Stammering with an Exposition of the Best methods of Cure, Medical, Surgical, and Educational*. London: Samuel Highley, 1843.

Poett, Joseph, *A Practical Treatise on Stammering and Nervous Affections of Speech, Illustrated with Several Cases of Cure, Authenticated by Persons of Distinction & Medical Authorities*. London: Calkin & Budd, 1833.

Prosser, James, *Lessons in Laryngoscopy Including Rhinoscopy and the Diagnosis and Treatment of Diseases of the Throat*. London: Baillier, Tindall and Cox, 1873.

Prosser, James, *Sore Throat, Its Nature, Varieties, and Treatment; Including the Connection Between Affections of the Throat and Other Diseases*. (3rd ed.). London: J. & A. Churchill, 1878.

Roe, W. R., *Poems on the Deaf and Dumb*. Derby: Francis Carter, 1888.

Sandlands, J. P. *The Voice and Public Speaking. A Book for All who Read and Speak in Public*. (2nd ed.). London: Hodder and Stoughton, 1880.

Schartz, Ch. Ed., *Des tumeurs du larynx*. Paris: Librairie J. B. Baillière et fils, 1886.

Schmauk, Theodore E., *The Voice in Speech and Song. A View of the Human Voice for Speakers and Singers and All Who Love the Arts of Speech and Song*. New York: John B. Alden, 1890.

Schrötter, L., *Vorlesungen über die Krankheiten des Kehlkopfes*. Vienna: Wilhelm Braumüller, 1892.

Sleight, William, *A Voice from the Dumb. A Memoir of John William Lashford*. London: Hamilton, Adams and Co., 1849.

Solly, J. Raymond, *Acting and the Art of Speech at the Paris Conservatoire. Hints on Reading, Reciting, Acting and the Cure of Stammering*. Elliot Stock, 1891.

Sterling Mackinlay, Malcolm, *Garcia the Centenarian and His Time; Being a Memoir of Manuel Garcia's Life and Labours for the Advancement of Music and Science*, London: Hutchinson & Co., 1908

Sutro, Emil, *Duality of Thought and Language. An Outline of Original Research*. (2nd ed.). London: Dryden House, 1904.

Tod, David, *The Anatomy and Physiology of the Organ of Hearing with Remarks on Congenital Deafness, the Diseases of the Ear, Some Imperfections of the Organ of Speech and the Proper Treatment of These Several Affections*. London: Longman, Rees, Orme, Brown, Green and Longman, 1832.

Treupel, Gustav, *Die Bewegungsstörungen im Kehlkopfe bei Hysterischen*. Jena: Verlag von Gustav Fischer, 1895.

Türck, Ludwig, *Praktische Anleitung zur Laryngoskopie. Mit 32 Holzschniten und einer Steindrucktafel*. Vienna: Wilhelm Braumüller, 1860.

Twistleton, Edward, *The Tongue Not Essential to Speech; with Illustrations of the Power of Speech in the African Confessors*. London: John Murray, 1873.

von Bruns, Victor, *Die erste Ausrottung eines Polypen in der Kehlkopfshöhle durch Zerschneiden ohne blutige Eröffnung der Luftwege nebst einer kurzen Anleitung zur Laryngoskopie.* Tübingen: Verlag der H. Laupp'schen Buchhandlung, 1862.

von Bruns, Victor, *Nachtrag zu meiner Schrift. Die erste Ausrottung eines Polypen in der Kehlkopfshöhle durch Zerschneiden ohne blutige Eröffnung der Luftwege.* Tübingen: Verlag der H. Laupp'schen Buchhandlung, 1863.

Von Meyer, G. H., *The Organs of Speech and Their Application in the Formation of Articulate Sounds.* London: Kegan Paul, Trench & Co., 1883.

Waxham, F. E., *Intubation of the Larynx.* Chicago: Charles Truax, 1888.

White, Harry W., *The Mechanism of Speech*, Manchester: s.l., 1897.

Wingrave, Wyatt, *Adenoids.* London: Baillier, Tindall and Cox, 1904.

Wright, Jonathan, *A History of Laryngology and Rhinology.* (2nd ed.). Philadelphia: Lea & Febiger, 1914.

Wyllie, John, *The Disorders of Speech.* Edinburgh: Oliver and Boyd, 1894.

Yearsley, James, *Stammering and Other Imperfections of Speech, Treated by Surgical Operations on the Throat.* London: Churchill, 1841.

Yearsley, Macleod, *Adenoids.* London: The Medical Times, 1901.

Yearsley, Walter A., *A Practical Self-Cure of Stammering and Stuttering.* Accrington: Mrs. Yearsley, 1909.

London: Royal College of Music

Adcock, John, *The Singer's Guide.* (5th ed.). London: J. Curwen & Sons, 1902.

Albani, Emma, *Forty Years of Song.* London: Mills & Boon, 1911.

Baker, J. Percy, *The Choir Boy's Handy Book of Musical Theory & Vocal Exercises.* London: "Musical News" Syndicate, 1898.

Barnett, John, *School for the Voice or the Principles of Singing, Being a Short Treatise Theoretical and Practical on the Culture and Development of the Voice, Consisting of a Series of Original & Progressive Exercises, Calculated to Facilitate & Perfect Execution Accor.* London: Addison & Hodson, s.d.

Behnke, Emil, *The Mechanism of the Human Voice.* (3rd orig.). London: J. Curwen & Sons, 1882.

Behnke, Mrs. Emil, [Kate Behnke] *The Speaking Voice: Its Development and Preservation.* (5th ed.). London: J. Curwen & Sons, 1897.

Bordogni, Marco, *Twenty-Four New Vocalises in a Progressive Form and Suitable for All Voices, Dedicated to His Majesty William III, King of Holland.* London: Boosey & Co., n.d.

Bristowe, John Syer, *The Physiological & Pathological Relations of the Voice and Speech.* London: David Bogue, 1880.

Buttz Clark, Felicia, *Guido the Choir-Boy.* London: Charles H. Kelly, 1906.

Caruso, Enrico, *How to Sing: Some Practical Hints.* London: John Church Company, 1914.

Cooke, Clifton, *Practical Singing. With Lessons on the Rendering of Their Songs by Claude Arundale, William J. Fenney, Gustave Ferrari [...] and Musical Illustrations and 19 Figures.* (2nd orig.). London: Kegan Paul, Trench & Co., 1910.

Curci, J., *Le petit solfège. Vingt solfèges pour voix de mezzo-soprano ou contralto avec accompaniment de piano.* London: Duncan Davison, s.d.

Cyril Johnson, the Choir Boy. A Tale. London: Simpkin, Marshall & Co., 1869.

Fétis, François Joseph, *Solfèges progressifs avec accompagnement de piano précédés des principes de la musique*. Paris: Maurice Schlesinger, s.d.

Garcia, Beata, *Hints on Singing by Manuel Garcia* (New & revi). London: Ascherberg, Hopwood and Crew Limited, 1911.

Garcia, Manuel, *Garcia's New Treatise on the Art of Singing. A Compendious Method of Instruction, with Examples and Exercises for the Cultivation of the Voice*. Cricklewood: Cramer, Beale, & Chappell, s.d.

Hamilton, *Hamilton's Modern Instructions in Singing. A Complete Compendium of the Rudiments of Music, Full Directions for the Cultivation and Management of the Voice, Numerous Exercises and Examples on Vocalization, with Solfeggi by Rossini and Sola and a Selection*. London: Robert Cocks and Co., 1853.

Haslam, W. E., *Style in Singing*. New York: G. Schirmer, 1911.

Hiles, Henry, *Fifteen Vocal Exercises*. London: Ashdown & Parry, s.d.

Holbrook Curtis, Henry, *Voice Building and Tone Placing. Showing a New Method of Relieving Injured Vocal Cords by Tone Exercises*. (2nd ed.). New York: D. Appleton and Company, 1909.

Ker Atkyns, Bellenden, *The Choir Master's Manual: Containing Full Instructions for Training a Choir, with Elementary Rules for Playing the Organ and Harmonium*. London: Addison and Lucas, s.d.

Lablache, Louis, *A Complete Method of Singing*. London: Chappell & Co., s.d.

Levien, John Mewburn, *The Singing Master's Decalogue. An Address Delivered Before the London Section of the Incorporated Society of Musicians on Saturday, 11th December, 1915*. London: Cary & Co., 1915.

Levien, John Mewburn, *Six Sovereigns of Song. Lectures*. London: Novello and Company, s.d.

Mann, Richard, *A Manual of Singing for the Use of Choir-Trainers and School-Masters, a Method of Learning to Sing at Sight*. London: Novello, Ewer and Co., s.d.

Masson, Elizabeth, *Vocal Exercises, Dedicated by Special Permission to H.SH. the Princess Ada of Hohenlohe Langenburg*, London, s.d.

Nava, G., *Metodo pratico di vocalizzazione per le voci di basso e baritono*. Milan: F. Lucca, s.d.

Rimbault, Edward F., *New Instructions on the Art of Singing; Comprising Directions for the Formation and Cultivation of the Voice, After the Methods of the Best Italian Master, Also a Series of Progressive Lessons, Studies in Vocalization, Solfeggio's etc, with an Accompanime*. London: Chappell & Co., s.d.

Roberts, J. Varley, *A Treatise on a Practical Method of Training Choristers*. (2nd ed.). London: Oxford University Press, 1900.

Röhner, G. W., *The Art of Singing Concisely and Fully Explained*. London: Robert Cocks and Co., s.d.

Rootham, Cyril Bradley, *Voice Training for Choirs and Schools*. Cambridge, UK: Cambridge University Press, 1912.

Shakespeare, William, *The Art of Singing. Based on the Principles of the Old Italian Singing-Masters, and Dealing with Breath-Control, Production of the Voice and Registers, Together with Exercises*. London: Metzler & Co., 1899.

Sims Reeves on "the Art of Singing." (2nd ed.). London: Chappell & Co., s.d.

Stockhausen, Julius, *Gesangsmethode*. Leipzig: C. F. Peters, 1884.

The Life of Miss Anne Catley, Celebrated Singing Performer of the Last Century Including an Account of Her Introduction to Public Life, Her Professional Engagements in London and Dublin and Her Various Adventures and Intrigues with Well-Known Men of Quality and Wealth, London, 1888.

Wiseman, Herbert, *School Choirs*. London: Paterson's publications, s.d.
Wood, Henry J., *The Gentle Art of Singing*. London: Humphrey Milford, 1927.
Zay, W. Henri, *Practical Psychology of Voice and of Life*. New York: G. Schirmer, 1917.

London: Royal Academy of Music

Bates, James, *Voice Culture*. London: Novello and Co., 1907.
Bishenden, Charles James, *The Voice and How to Use It. Describing the Mechanism of the Vocal Organs, How to Strengthen the Lung, etc.* London: J. Miles & Co., 1869.
Breare, W. H., *Vocal Faults and Their Remedies*. London: Simpkin, Marshall, Hamilton, Kent & Co., 1907.
Browne, Lennox, *Medical Hints on the Production and Management of the Singing Voice*. London: Chappell & Co., 1877.
Curwen, J. Spencer, *Studies in Worship—Music, Chiefly as Regards Congregational Singing*. London: J. Curwen & Sons, 1880.
Davidson Palmer, E., *The Rightly Produced Voice. A Presentation of Facts and Arguments in Support of a New Theory of Voice-Production*. London: Joseph Williams Ltd., 1909.
De la Madelaine, Stephen, *Théories complètes du chant*. Paris: Amyot, 1852.
Dunn, Sinclair, *The Solo Singer.* (9th ed.). London: J. Curwen & Sons, 1909.
Ely, Thomas, *The Elements of Voice-Production and Singing*. London: Breitkopf & Härtel, 1906.
Estcourt, Mary Jane, *Music the Voice of Harmony in Creation*. London: Longman, Brown, Green and Longmans, 1857.
Evans, John, and McNaught, W. G., *The School Music Teacher. A Guide to Teaching Singing in Schools by Tonic Sol-Fa Notation and Staff Notation*. (9th ed.). London: J. Curwen & Sons, 1903.
Ffrangcon-Davies, D. *The Singing of the Future*. London: John Lane, 1904.
Garcia, Gustave, *The Singing Teacher's Notebook*. London: Boosey & Co., 1910.
Gilbert, Alfred, *A Singing Book for Public School and Choir Boys*. London: Novello and Company, s.d.
Hayle Walshe, Walter, *Dramatic Singing, Physiologically Estimated*. London: Kegan Paul, Trench & Co., 1881.
Heinrich, Max, *Correct Principles of Classical Singing*. Boston: Lothrop, Lee & Shepard Co., s.d.
Henderson, W. J., *The Art of the Singer*. London: John Murray, 1906.
Holmes, Gordon, *The Science of Voice Production and Voice Preservation for the Use of Speakers and Singers*. London: Chatto & Windus, 1880.
Hulbert, H. H., *Breathing for Voice Production*. London: Novello and Company, 1903.
Hullah, John, *Wilhem's Method of Teaching Singing, Adapted to English Use*. London: John W. Parker, 1842.
Joal, Joseph, *On Respiration in Singing*. London: F. J. Rebman, 1895.
Kelly, Thomas, *The First Principles of Voice Production in Song and Speech*. London: Burns and Oates, Ltd., 1893.
Lawton, William H. *The Singing Voice and Its Practical Cultivation*. New York: Charles Francis Press, 1901.
Lunn, Charles, *Artistic Voice in Speech and Song*. London: Baillier, Tindall and Cox, 1884.
Lunn, Charles, *Philosophy of Voice, Showing the Right and Wrong Action of Voice in Speech and Song.* (9th ed.). London: Baillier, Tindall and Cox, 1900.
Lunn, Charles, *The Voice: Its downfall, Its Training and Its Use. A Manual for Teachers, Singers, and Students*. London: Reynolds & Co., 1904.

Myer, Edmund J. *The Renaissance of the Vocal Art. A Practical Study of Vitality, Vitalized Energy, of the Physical, Mental and Emotional Powers of the Singer, Through Flexible, Elastic Bodily Movements*. Boston: Boston Music Company, 1902.

Nares, James, *A Treatise on Singing, in Which Is Exhibited and Explained by Examples All the Known Rules of Solmisations, or Learning to Sing by Notes, with Directions for the Delivery and Management of the Voice, Drawn from Observations on Nature as Well as Art*, London: Welcker, s.d.

Rowley, Christopher E., *The Voice: Or the Physiologist Versus the Singing Master*. London: Office of Music, 1898.

Rush, James, *The Philosophy of the Human Voice, Embracing Its Physiological History; Together with a System of Principles, by Which Criticism in the Art of Elocution May Be Rendered Intelligible, and Instruction, Definite and Comprehensive*. (3rd ed.) Philadelphia: J. Crissy, 1845.

San Carolo, Irene, and Daniel, Patrick, *The Common-Sense of Voice Development*. London: Baillier, Tindall and Cox, 1906.

Seiler, Emma, *The Voice in Singing*. Philadelphia: J. B. Lippincott Company, 1893.

Stephenson, Robert, *How to Breathe, Speak and Sing*. London: Jarrold & Sons, s.d.

Sterling Mackinlay, Malcolm, *The Singing Voice and Its Training*. London: George Roútledge & Sons, 1910.

Taylor, David C., *The Psychology of Singing. A Rational Method of Voice Culture Based on a Scientific Analysis of All Systems, Ancient and Modern*. New York: Macmillan and Co., 1908.

Wood, Henry J., *To my Pupils*. London: for private circulation only, 1896.

Ziska, Leopoldine, *Handbook of the Four Elements of Vocalization*. London: Ewer & Co., 1850.

Paris: National Library

B., A. de, Les jeunes cantatrices. *Gazette Musicale de Paris*, 40:2 (1835), 321–323.

Ballande, Hilarion, *La parole, ou l'art de dire et d'exprimer, appliqué à la causerie, au professorat, à la lecture à haute voix, au barreau, à la scène, à la tribune, et à la chaire sacrée*. Paris: E. Dentu, 1868.

Batiste, Eduoard, *Petit solfège harmonique a la portée des plus jeunes voix*. Paris: Henri Heugel, 1868.

Beaux, Jean-Jacques, *De l'influence de la magnétisation sur le développement de la voix et du goût en musique*. Paris: Edouard Garnot, 1855.

Bennati, François, *Recherches sur le mécanisme de la voix humaine*. Paris: J.-B. Baillière, 1832.

Bennati, François, *Recherches sur les maladies qui affectent les organes de la voix humaine*. Paris: J.-B. Baillière, 1832.

Bennati, François, Du mécanisme de la voix pendant le chant. *Revue Musicale*, 8 (1830), 353–375.

Bernard, Joseph-Ferdinand, *Gymnastique pulmonaire, ou l'art de respirer pour chanter, parler et vivre*. Lyon: impr. de Jevain et Bourgeon, 1868.

Blanc, Honoré, *Okygraphie, ou l'art de fixer, par écrit, tous les sons de la parole*. Paris: Locard et Davi, 1819.

Blanchard, Henri, Des bains orientaux et russes, ou physiologie du chanteur. *Gazette Musicale de Paris*, 8:35, 1841.

Boisquet, M. F., *Essais sur l'art du comédien chanteur*. Paris: Longchamps, 1842.

Bonheur, Georges, *Exercices pratiques de la voix et de la respiration*. Paris: P. Dupont, 1891.

Bouché, L., *De L'Art du chant, théorie nouvelle basée sur l'appréciation des éléments constitutifs de la voix*. Nogent-le-Rotrou: Imprimerie de A. Gouverneur, 1872.

Bouffé, Floris, *De l'influence du liquide organique (extractum testis) dans les affections des voies aériennes supérieures: ses bons effets sur la voix.* Clerment (Oise): Imprimerie de daix frères, 1895.

Bourdy, Lucien, *Cours pour la cure du bégaiement. Rapport à la Société de médecine de la Sarthe sur la méthode de Chervin, par une commission officielle.* Le Mans: impr. de A. Leguicheux, 1872.

Boyer, August, *De la mue de la voix chez le jeune sourd parlant: thèse pour l'aggrégation de l'enseignement des sourds-muets présentée et soutenue en octobre 1892.* Paris: G. Carré et C. Naud, 1893.

Boyer, August, *Comment on peut apprendre à parler à un jeune semi-idiot non sourd-muet de naissance.* Paris: G. Carré et C. Naud, 1896.

Broster, John, *Progrès de la méthode pour la cure efficace des embarras de la parole.* Paris: impr. de C. Doyen, 1826.

C. H., *Petite théorie musicale des enfants à chanter sur quelques airs connus.* Paris: Prosper Pégiel, 1874.

Canton, F., *Cours complet de sténographie pratique s'apprenant sans maître et permettant de suivre la parole.* Bordeaux: impr. de G. Delmas, 1894.

Carillon, Louis, *Traitement nouveau des maladies de la gorge, du nez, du larynx (voix) et des oreilles (surdité),* Paris, 1897.

Carrier, Albert, *Etude sur la localisation dans le cerveau de la faculté du langage articulé.* Paris: G. Baillière, 1897.

Castex, André, *Hygiène de la voix parlée et chantée.* Paris: Masson, 1894.

Castex, André, *Maladies de la voix.* Moscow: impr. de I. N. Kouchnerev, 1898.

Castex, André, *Maladies du larynx, du nez et des oreilles.* Paris: J.-B. Baillière, 1899.

Castex, André, *L'orthophonie, correction des troubles de la parole et de la voix: conférence faite à la Sorbonne.* Paris: Institution nationale des sourds-muets, 1905.

Chabbert, Louis, *Deux cas de bégaiement hystérique chez des dégénérés.* Paris: L. Battaille et Cie, 1893.

Chassan, Joseph-Pierre, *Traité des délits et contraventions de la parole, de l'écriture et de la presse.* Paris: Videcoq, 1837.

Chervin, Amedée, *Statistique des bègues pour le département de la Gironde, de 1851 à 1870, suivie de l'exposé pratique de la méthode préservative et curative du bégaiement.* Bordeaux: impr. de G. Gounouilhou, 1874.

Chervin, Arthur, *Exercices de lecture à haute voix et de récitation, divisions élémentaires: prononciation française.* Paris: Chervin, 1880.

Chervin, Arthur, *La méthode Chervin, pour corriger le bégaiement et tous les autres défauts de prononciation: expliquée dans 38 rapports officiellement rédigés par l'Académie de médecine, le Conseil de santé des armées, les sociétés médicales de Lyon, Marseille, Toulouse.* Paris: Chervin, 1881.

Chervin, Arthur, *Bégaiement et autres défauts de prononciation.* Paris: Société d'éditions scientifiques, 1891.

Chervin, Arthur, *Bégaiement et autres maladies fonctionnelles de la parole.* Paris: Société d'éditions scientifiques, 1902.

Chervin, Claudius, *Méthode-Chervin appliquée à la cure du bégaiement et de tous les autres défauts de prononciation, méthode autorisée et recommandée par M. le ministre de l'Instruction publique.* Paris: Institution des bègues, 1872.

Collet, N., *Solfèges de l'orphéon et des écoles. Recueil de 333 leçons à deux voix égales.* Paris: Garnier frères, 1868.

Colombat, Emile, *Cours d'orthophonie, annexé à l'Institution nationale des sourds-muets, pour le redressement du bégaiement et des vices de la parole.* Paris: impr. de Noblet, 1875.

Colombat de l'Isère, Marc, *Tableau synoptique et statistique de toutes les espèces de bégaiement : et des moyens curatifs qui conviennent à chaque variété.* Paris: Chez les principaux libraires de médecine, 1833.

Colombat de l'Isère, Marc, *Le mécanisme des cris et leur intonation notée dans chaque espèce de douleurs physiques et morales*. Paris: impr. de Moquet, 1840.

Colombat de l'Isere, Marc *Réponse au Dr Becquerel, ou Réfutation de son traité du bégaiement*. Paris: Chez tous les libraires de médecine, 1844.

Colombat de l'Isere, Marc, *L'enseignement ortophonique au conservatoire national de musique et de déclamation de Paris*. Paris: P. Asselin, 1873.

Costa, Andrea, *Considerations on the Art of Singing in General*. London: Schulze, Poland Street, 1824.

Coulange, D. A. P., *Écriture des sons, ou L'Art d'écrire toutes les langues presque aussi vite que la parole*. Paris: Chez l'auteur, 1857.

Coupey, Alphonse Arthur, *Maladies de la voix: prophylaxie et traitement*. Le Mans: imprimerie de E. Monnoyer, 1892.

Cresp, Joseph, *Essai sur la déclamation oratoire et dramatique, la diction et la prononciation ; suivi d'une Nouvelle méthode curative du bégaiement et de tous les vices de la parole*. Paris: Hachette, 1837.

Crosilhes, Hippolyte-Pierre, *Hygiène et maladies de la poitrine et de la voix*. Paris: Moquet, 1847.

Crosti, Eugène, *La voix des enfants*. Paris: Imprimerie Brevetée Charles Blot, 1881.

Cuvier, M., and Savart, Rapport sur un mémoir de M. Bennatti. *Revue Musicale*, 8 (1830), 33–40.

Dax, Marc, "Lésion de la moitié gauche de l'encéphale coïncidant avec l'oubli des signes de la pensée," *Gazette hebdomadaire de medicine et de chirurgie*, 17, 1865.

De Fleury, Armand, *Des tentatives de localisation de la parole d'un seul côté du cerveau : pourquoi la lésion paraît se rencontrer plus fréquemment dans le lobe antérieur gauche que dans le droit*, s.l., 1866.

De la Cour, Emile, *Lectures classiques, spécialement destinées aux écoles primaires et supérieures . . . par Émile de La Cour, et précédées d'un enseignement raisonné sur la lecture à haute voix, par Ed. Mennechet*. Paris: Delloye, 1840.

De Rillé, Laurent, *Du chant choral*. Paris: Perrotin, editeur de l'orpheon, 1856.

Debay, Auguste, *Hygiène de la voix et gymnastique des organes vocaux*. Paris: Moquet, 1852.

Defer, E., *Examen du strabisme et du bégaiement*. Paris: Lucas, 1841.

Dekytspotter, Emile, *Alphabet intellectuel, ou Esprit des lettres, physiologie de la parole, clef de la science*. Lille: Imprimerie de Horemans, 1867.

Deleschamps, Albert, *Etude physique des sons de la parole*. Paris: Savy, 1869.

Despiney, Félix, *Physiologie de la voix et du chant*. Paris: J.-B. Baillière, 1841.

Diday, M., and Pétrequin, M., *Mémoire sur une nouvelle espèce de voix chantée*. Félix Malteste, 1840.

Dieffenbach, Johann Friedrich, *Lettre adressée à l'Institut de France sur la guérison du bégaiement au moyen d'une nouvelle opération chirurgicale*. Berlin: A. Foerstner, 1841.

Dorval, A., *L'art de laproniciation appliquée au chant*. Paris: Chez l'auteur, 1866.

du Collet, Cléricy, *La voix amplifiée ou recouvrée par l'éducation ou la réeducation des muscles du larynx*. Paris: impr. de Chamerot et Renouard, 1899.

Dubroca, Louis, *Principes raisonnés sur l'art de lire à haute voix, suivis de leur application particulière à la lecture des ouvrages d'éloquence et de poésie*. Paris: Imprimerie de Farge, 1802.

Du faucet, ou voix pharyngienne. *Revue Musicale*, 27 (1835) 212–214, 217–219.

Duployé, Emile, *Sténographie Duployé, ou L'Art de suivre avec l'écriture la parole la plus rapide appris sans maître en quelques heures*. Aisne: Duployé, 1867.

Duprez, Gilbert, *Souvenirs d'un chanteur*. Paris: Callman Lévy, 1880.

Duprez, Gilbert, *Joyeusetés d'un chanteur dramatique*. Paris: Tresse, s.d.

Duquesnois, Jean, *Manuel de l'orateur et du lecteur*. Paris: Chez Vaton, 1841.

Expériences sur la voix faites par mm. Bennati, Savart et Gagnard-de-Latour, *Revue Musicale*, 45 (1831), 357–359.

F., Du timbre et du mordant de la voix, de la parole et du chant pendant l'inspiration. *Revue Musicale*, 7 (1830) 49–51.

Fauré, Jean, *La voix et le chant. Traité pratique*. Paris: Henri Heugel, 1886.

Faure-Lacroze, S., *Essai sur la composition d'un nouvel alphabet pour servir à représenter les sons de la voix humaine*. Paris: Firmin Didot frères, fils et cie, 1831.

Fétis, Eduoard, Physiologies musicales: le chânteur de salon. *Gazette Musicale de Paris*, 21:4 (1854), 26–28.

Fétis, Père, Des divers systèmes concernant la voix humaine et par occasion, des causes présumables de la dégénération de cet organe à l'époque actuelle. *Gazette Musicale de Paris*, 16:24 (1849).

Fétis, Revue critique. Methode élémentaire et progressive de chant, contenant une collection de musique sacrée ancienne et moderne, etc. par Francois Stoepel. *Gazette Musicale de Paris*, 2:52 (1835), 427–428.

Fourcade, V. M., *De la Parole, du sens de l'ouïe et des sourds-muets*. Caen: Imprimerie de Domin, 1858.

Gaillard, J., *Lecture et prononciation d'après une méthode entièrement nouvelle . . . basée sur le jeu des organes de la parole et pouvant servir à l'enseignement simultané des entendants parlants et des sourds-muets*. Paris: Delagrave, 1868.

Garnault, Paul, *Cours théorique et pratique de physiologie, d'hygiène et de thérapeutique de la voix parlée et chantée*. Paris: A. Maloine, 1896.

Gerber, Gustav, *Die Sprache als Kunst*. Bromberg: H. Heyfelder, 1871.

Gerdy and Bennati, Des réclamations de M. le docteur Gerdy à propos du mémoire de M. Bennati. *Revue Musicale*, 8 (1830), 161–169.

Gobin, Henry, *L'art de peindre la parole : études sur l'imprimerie, la librairie, les cartes et globes, la fonderie en caractères, la stéréotypie, la polytypie, la lithographie, la gravure sur bois, sur cuivre, sur pierre, etc*. Paris: E. Lacroix, 1874.

Godard, Jules, *Du Bégaiement et de son traitement physiologique*. Paris: J. B. Baillière et fils, 1877.

Gouguenheim, Achille, and Lermoyez, Marcel, *Physiologie de la voix et du chant: hygiène du chanteur*. Paris: A. Delahaye et E. Lecroisnier, 1885.

Grimon, Benjamin, *Méthode curative du bégaiement, méthode Chervin, rapport à M. le recteur de l'Académie de Douai*, Lille, 1875.

Grognier-Quélus, Jean-Baptiste, *Etudes sur l'art dramatique et oratoire : conseils aux comédiens & comédiens chanteurs*. Brussels: impr. de Detrie-Tomson, 1858.

Grosselin, Augustin, *Alphabet phonomimique*. Corbeil: impr. de Crété et fils, 1864.

Grosselin, Augustin, *Manuel de la phonomimie, ou Méthode d'enseignement par la voix et par le geste*. Paris: Bélin frères, 1905.

Gubian, Louis, *Rapport de la commission spéciale instituée par M. le sénateur préfet du Rhône pour l'examen de la méthode curative du bégaiement*. Lyon: impr. de A. Vingtrinier, 1866.

Guéroult, Adrien, Cours de chant et de tenue dramatique. *Gazette Musicale de Paris*, 6:42 (1839), 332–334.

Guéroult, Adrien, De l'enseignement populaire du chant. Réunion générale des cinq divisions de l'orphéon. *Gazette Musicale de Paris*, 6:18 (1839), 137–139.

Guillaume, Adrien, *Du Bégaiement et de son traitement (résumé)*. Paris: J.-B. Baillière, 1872.

Guillemin, Auguste, *Notions d'acoustique. Introduction à l'étude de la phonation*. Paris: Société d'éditions scientifiques, 1895.

Habay, F., *Unité de la voix. Méthode synthétique du chant et de la parole.* Paris: Ancienne maison Quantin, 1894.

Haeck, Henry, *Méthode de musique vocale pratique, théorique et pédagogique suivie de l'art de chanter en choeur.* Paris: Belin frères, 1890.

Hamonic, P., and Schvartz, E., *Manuel du chanteur et du professeur de chant.* Paris: Librairie Fischbacher, 1888.

Hennelle, Claire, *Rudiment des chanteurs.* Paris: J. Meissonier, 1843.

Jarry, J., *Enseignement du chant dans les écoles.* Lille: L. Quarré, 1873.

Jourdant, G. L., *De la Guérison du bégaiement, par Jourdant, méthode nouvelle présentée à l'Institut (Académie des sciences), dans sa séance du 12 juin 1843.* Paris: impr. de A. Appert, 1843.

Jourdant, G. L., *Lettre de M. Jourdant à MM. les membres de l'Académie de médecine.* Paris: impr. de A. Appert, 1843.

Jumel, A. (n.d.), *Notions Principes & Exercices pour l'étude quotidienne du chant (spécialement écrits pour la voix de femme).* Bureau d'édition de la schola Cantorum.

Jutet, Jean-Etienne, *Du Bégaiement et de sa guérison, extraits du rapport à MM. les membres du jury à l'Exposition universelle de Lyon 1873 sur la méthode de Chervin, par le comité pour l'instruction publique.* Lyon: impr. de Schneider frères, 1873.

La Fage, Adrien, Mémoire sur la voix humaine. *Gazette Musicale de Paris*, 14:29 (1847).

Laisné, Marius, *L'action, comprenant l'animation de la voix, de la physionomie et du geste.* Paris: Chez l'auteur, 1867.

Laisné, N., *Recueil de chants spéciaux notés et numérotés pour rythme et exercices gymnastiques et fortifier la voix ainsi que tous les organes qui concourent à son action.* Paris: Librairie classique Picard-Bernheim, 1882.

L'art de chanter agréablement enseignement en 20 leçons. Paris: au cabinet littéraire, chez barbe, 1831.

Laugel, Auguste, *La voix, l'oreille et la musique.* Paris: G. Baillière, 1867.

Le Maout, Emmanuel, *Leçons analytiques de lecture à haute voix.* Paris: E. Ducrocq, 1852.

Lefort, Jules, *L'émission de la voix chantée.* Paris: Lemoine & Fils, 1893.

Lemoine, Albert, *De la physionomie et de la parole.* Paris: G. Baillière, 1865.

Letellier, Charles-Louis-Augustin, *Les lois de la parole, ou Examen critique des bases sur lesquelles reposent les langues orientales et occidentales, mortes et vivantes.* Paris: B. Duprat, 1861.

Luneau, Louis-Henri, *Du Traitement des bègues par la méthode Chervin, rapport adressé à M. le maire de Nîmes.* Nîmes: impr. de Roger et Laporte, 1873.

M. *Actualité, par M*** récitée dans ses séances par M. V.-M. Fourcade, artiste dramatique et lyrique, maître en l'art de parler, enseigneur de la parole aux sourds-muets, professeur de prononciation, correcteur de tout vice d'articulation et de toute défectu.* imp. de A. Loger, Le Mans: C.-J. Boulay et Cie, 1861.

Magnat, Marius, *Cours d'articulation, enseignement de la parole articulée aux sourds-muets.* Paris: Sandoz et Fischbacher, 1874.

Mainzer, Joseph, Enseignement du chant dans les écoles élémentaires. *Gazette Musicale de Paris*, 2:22 (1835), 185.

Mainzer, Joseph, *Méthode de chant pour les enfants à l'usage des écoles et des pensions.* Paris: chez maurice schlesinger, 1836.

Mainzer, Joseph, *Méthode de chant pour voix d'hommes à l'usage des collèges, écoles normales, écoles militaires et cours d'ouvriers.* Paris: au dépot des publications Mainzer, 1838.

Mainzer, Joseph, *École chorale contenant la grammaire musicale, la théorie des accords, le contrepoint, l'imitation, la fugue, etc.* Paris: Chez Tantenstein, 1839.

Malebouche, Fortuné, *Institution de M. Malebouche. Nouvelle méthode pour guérir le bégaiement*. Paris: impr. de Ducessois, 1828.

Malebouche, Fortuné, *Précis sur les causes du bégaiement, et sur les moyens de le guérir*. Paris: Fortin, Masson, 1841.

Mandl, Louis, *Cours d'hygiène de la voix professé au Conservatoire de musique*. Paris: Imprimerie de Alcan-Lévy, 1873.

Mandl, Louis, *Hygiène de la voix parlée ou chantée, suivie du formulaire pour le traitement de la voix*. Paris: J.-B. Baillière, 1876.

Marage, René, *Développement de l'énergie de la voix par des exercices respiratoires*. Paris: Chez l'auteur, 1907.

Martin, Julien, De l'amélioration du chant dans les collèges oryaux de paris. *Gazette Musicale de Paris*, 12:41 (1845), 337–339.

Martin, Julien, Le violoncelle et la voix humaine. *Gazette Musicale de Paris*, 12:49 (1845), 397–400.

Masson, A. [Antoine Filibert], and Longet [F.A.], *Études expérimentales sur la voix et sur les causes de la production du son dans divers instruments de musique*. Paris: Victor Masson, 1852.

Mathieu, Honoré, *De la parole et du bégaiement*. Paris: Chez l'auteur, 1847.

Mennechet, Edouard, *Etudes sur la lecture à haute voix*. Paris: Langlois et Leclerq, 1855.

Miquel-Chaudesaigues, Mme., *Leçons nouvelles sur l'art vocal*. Paris: Librairie de Firmin-Didot et cie, 1880.

Morin, Laurent-Joseph, *Traité de prononciation, indiquant les moyens d'obtenir une bonne émission de voix, de corriger les accents vicieux et tous les défauts de prononciation*. Paris: Passage Saulnier, 1852.

Moulin, Etienne, *Quelques mots sur la respiration, la voix et la musique*. Paris: Imprimerie de J. Claye, 1865.

Netter, Abraham, *La parole intérieure et l'âme*. Paris: Berger-Levrault, 1892.

Nouvelles: cloche Tabarie. *Gazette Musicale de Paris*, 6:5 (1839), 46.

Paignon, Eugène, *Eloquence et improvisation: art de la parole oratoire au barreau, à la tribune, à la chaire*. Paris: Cotillon, 1846.

Panofka, Heinrich, *Abécédaire vocal, méthode préparatoire de chant*. Paris: G. Brandus, Dufour et cie, 1858.

Passot, Philippe-Claude, *Rapport présenté à la Société d'éducation de Lyon, par une commission désignée par elle et composée de M. le Dr Desgranges, . . . MM. les Drs Fonteret et Passot . . . sur la méthode employée pour la cure du bégaiement et de tous les autres vices de prononciatio*. Lyon: impr. d'A. Vingtrinier, 1864.

Pavid, Henriette, *Traité sur la guérison du bégaiement et de toutes les défectuosités de prononciation*. Paris: Durand, 1864.

Perini, Carlo, *Méthode pour enseigner par la parole la langue maternelle aux sourds-muets*. Currière: Imprimerie de l'école des sourds-muets, 1890.

Pétrequin, M., and Diday, M. Mémoire sur le mécanisme de la voix de fausset. *Gazette Médicale de Paris*, 11: 12 (1843), 4–21.

Potton, A. *Observations physiologiques et musicales. Analyse d'un mémoire des docteurs Diday et Petrequin*. Lyon: Imprimerie de L. Boitel, 1843.

Prévost, Hyppolyte, *Nouveau manuel de sténographie ou l'art de suivre la parole en écrivant*. Paris: Roret, 1834.

Profumo, Louis, *De l'organe de la parole interne*, Paris: Imprimerie de Pillet, 1860.

Ramont, Mansuès-François, *De la voix et de la parole*. Paris: Imprimerie de Feugueray, 1803.

Riant, Aimé, *Hygiène de l'orateur*. Paris: J. B. Baillière et fils, 1886.

Rivail, Charles, *De l'art de la parole au barreau, disocurs prononcé à la séance solennelle des conférences.* Grenoble: impr. de F. Allier père et fils, 1890.

Rouma, Georges, *La parole et les troubles de la parole.* Paris: H. Paulin, 1907.

Saint-Michel, Jean-Baptiste S. *Discours sur les voyelles, les adjectifs, les interjections, le verbe et sur l'exécution de la loi de l'épargne d'action dans l'exercice de la parole.* Strasburg: Levrault, 1805.

Sandras, A., *La voix modifiée par les inhalations.* Paris: G. Masson, 1894.

Segond, Dr., *Mémoire sur la parole.* Paris: Rignoux, 1847.

Segond, Dr., *Mémoire sur la voix inspiratoire.* Paris: Rignoux, 1848.

Segond, Dr., *Mémoire sur les modification du timbre de la voix humaine.* Paris: Rignoux, 1848.

Segond, Dr., *Note sur les mouvements de totalité du larynx.* Paris: Rignoux, 1848.

Segond, L. A., *Mémoire sur l'ossification des cartilages du larynx.* Paris: Rignoux, 1847.

Surville, Clovis, *Guérison du bégaiement, exposé d'une nouvelle méthode.* Toulouse: F. Gimet, 1872.

Sténographie manuelle. Au moyen de cette nouvelle méthode aussi rapide que la parole, on peut tenir conversation secrète en public, et se faire parfaitement comprendre des sourds muets. Paris: lith. de Leclerq et Chevalier, 1858.

Torramoréll, Michel-Bonavonture-François and Clement, Félix, *Méthode de musique vocale graduée et concertante pour apprendre à solfier et à chanter a une et plusieurs voix.* Paris: Firmin Didot frères, fils et cie, 1871.

Trousseau, A., and Belloc, H., *Traité pratique de la phtisie laryngée, de la laryngite chronique et des maladies de la voix.* Paris: J.-B. Baillière, 1837.

Vacher, L., (1877). *De la voix chez l'homme: au point de vue de sa formation, de son étendue et de ses registres.* G. Masson.

Vernet, Claude, and Vernet, Coralie, *Leçons théoriques et pratiques sur la cause, le siège, la nature, le mécanisme et le traitement du bégaiement, ou Méthode naturelle de traitement du bégaiement en 20 leçons, mise à la portée de tout le monde par Mlle Coralie Vernet et Claude Vernet son père.* Clerment-Ferrand: A. Veysset, 1841.

Violette, Cyrille Florimond, *Études sur la parole et ses défauts et en particulier du bégayement.* Paris: Leclerc, 1862.

Voisin, Félix, *Du Bégaiement, ses causes, ses différens degrés, influence des passions, des sexes, des âges, etc. sur ce vice de prononciation ; moyens thérapeutiques pour prévenir, modifier ou guérir cette infirmité.* Paris: Croullebois, l'auteur, 1821.

Wilhorgne, Charlemagne, *Dactylographie, ou Sténographie des doigts, art de converser avec les sourds-muets instruits au moyen de ce procédé, avec une vitesse qui rivalise presque avec la parole.* Rouen: Imprimerie de Surville, 1846.

Paris: Library Hector Berlioz

Bayard Potron, Charles, and Boulanger, Ernest, *Une voix. Opéra comique en un acte.* Paris: Beck, 1845.

Berlioz, Hector, *A travers chants.* Paris: Grund, 1971.

Fétis, François Joseph, *La musique mise à la portée de tout le monde.* Paris: Chez Paulin, 1836.

Gaugiran-Nanteuil, P. Charles and Berton, Henri, *Le charme de la voix. Opéra comique en un acte.* Paris: Chez Barba, 1812.

Gérard, Henri-Philippe, *Considérations sur la musique en général, et particulièrement sur tout ce qui a rapport a la vocale.* Paris: Kleffer, 1819.

Giraud, J. Fréderic, *Le polycorde ou nouvelle méthode théorique et pratique de musique vocale et de musique instrumentale.* Paris: Librarie Adrien Le Clerc, 1869.

Mélesville, [Duveyrier, Anne-Honoré-Joseph] and Alary, G., *La voix humaine. Opéra en deux actes.* Paris: Madame Jonas, Libraire de l'Opéra, 1861.

Parino, Marcello, *Nouvelle méthode de chant.* Paris: ébrard, s.d.

Tiersot, Julien, *Programme d'un recueil de chants à l'usage des écoles primaires de France.* Paris: Librairie Hachette et cie, 1893.

SECONDARY READING

Abbate, Carolyn, *Unsung Voices. Opera and Musical Narrative in the Nineteenth Century*, Princeton, NJ: Princeton University Press, 1991.

Abberley, Will, "Race and species essentialism in nineteenth-century philology," *Critical Quarterly*, 3:4 (2011) pp. 45–60.

Alberti, Fay Bound, *Matters of the Heart: History, Medicine and Emotion*, Oxford: Oxford University Press, 2010.

Alter, Stephen G., *Darwinism and the Linguistic Image, Language, Race and Natural Theology in the Nineteenth Century*, Baltimore: Johns Hopkins University Press, 2003.

André, Naomi, *Black Opera: History, Power, Engagement*, Champaign: University of Illinois Press, 2018.

André, Naomi, *Voicing Gender. Castrati, Travesti, and the Second Woman in Early-Nineteenth-Century Italian Opera*, Bloomington: Indiana University Press, 2006.

Asseraf, Arthur, *Electric News in Colonial Algeria*, Oxford: Oxford University Press, 2019;

Austin, Stephen F. "'Like the Squawk of a Capon'—the Tenor Do Di Petto," *Journal of Singing*, 61:3 (2005).

Barbier, Patrick, *The World of the Castrati: The History of an Extraordinary Operatic Phenomenon*, London: Souvenir Press, 1998.

Barthes, Roland, "Le grain de la voix," *L'obvie et l'obtus*, Paris: Seuil, 1982.

Barthes, Roland, "The Grain of the Voice," Stephen Heath (ed.), *Image, Music, Text,* London: Fontana Press, 1977.

Bateman, Chimène, "Irrepressible Malebouche. Voice, Citation and Polyphony in the Roman de la rose," *Cahiers de recherches médiévales et humanistes*, 22 (2011), pp. 9–22.

Bederman, Gail, *Manliness and Civilization. A Cultural History of Gender and Race in the United States, 1880–1917*, Chicago: University of Chicago Press, 1995.

Bergeron, Katherine, *Voice Lessons. French Mélodie in the Belle Epoque*, Oxford: Oxford University Press, 2010.

Bijsterveld, Karin, *Sonic Skills: Listening for Knowledge in Science, Medicine and Engineering*, London: Palgrave Macmillan, 2019.

Bloom, Gina, *Voice in Motion: Staging Gender, Shaping Sound in Early Modern England*, Philadelphia: University of Pennsylvania Press, 2007.

Boddice, Rob, and Mark Smith, *Emotion, Sense, Experience*, Cambridge, UK: Cambridge University Press, 2020.

Boddice, Rob, "Vivisecting Major: A Victorian Gentleman Scientist Defends Animal Experimentation, 1876–1885," *Isis*, 102:2 (2011), pp. 215–237.

Boddice, Rob, *The Science of Sympathy: Morality, Evolution, and Victorian Civilization*, Champaign: University of Illinois Press, 2016.

Boehrer, Bruce Thomas, *Parrot Culture: Our 2500-Year-Long Fascination with the World's Most Talkative Bird*, Philadelphia: University of Pennsylvania Press, 2015.

Bourke, Joanna, *What It Means to Be Human: Reflections from 1791 to the Present*, London: Counterpoint, 2011.

Boyd, S., and N. Hewlett, "The Gender Imbalance Among Speech and Language Therapists and Students," *International Journal of Language and Communication Disorders*, 36 (2001) pp. 167–172.

Brett, Philip, Elizabeth Wood, and Gary Thomas, *Queering the Pitch: The New Gay and Lesbian Musicology*, New York: Routledge, 1994.

Broomhall, Susan, *Early Modern Emotions: An Introduction*, New York: Routledge, 2017.

Brown, Michael, *Performing Medicine: Medical Culture and Identity in Provincial England, c. 1760–1850*, Manchester: Manchester University Press, 2011.

Bruyninckx, Joeri, *Listening in the Field: Recording and the Science of Birdsong*, Cambridge, MA: MIT Press, 2018.

Burke, Peter, *The Art of Conversation*, London: Polity, 1993.

Buscatto, Marie, *Femmes du jazz*, Paris: Editions CNRS, 2018.

Butler, Judith, *Frames of War: When is Life Grievable?*, London: Verso Books, 2009.

Butler, Shane, *The Ancient Phonograph*, New York: Zone Books, 2015.

Cavarero, Adriana, *For More Than One Voice: Toward a Philosophy of Vocal Expression*, Stanford, CA: Stanford University Press, 2005.

Chapman, David, *Sandow the Magnificent: Eugen Sandow and the Beginnings of Bodybuilding*, Champaign: University of Illinois Press, 2006.

Chiss, Jean Louis, "Les linguistes du XIXe siècle, l'«identité nationale» et la question de la langue," *Langages*, 182:2 (2011), pp. 41–53.

Cleall, Esme, "Deafness: Representation, Sign Language, and Community, ca. 1800–1920," in: Joyce L. Huff and Martha Stoddard Holmes (eds.), *A Cultural History of Disability in the Long Nineteenth Century*, London: Bloomsbury, 2020, pp. 95–112.

Cleall, Esme, "Decolonising Deaf History. Harlan Lane, Postcolonialism, and Critical Colonial History," *Zeithistorische Forschungen/Studies in Contemporary History*, 19 (2022), pp. 380–387.

Connor, Stephen, *Dumbstruck: A Cultural History of Ventriloquism*, Oxford: Oxford University Press, 2000.

Connor, Steven, *Beyond Words: Sobs, Hums, Stutters and Other Vocalizations*, Chicago: University of Chicago Press, 2014.

Cook, Nicholas, *Beyond the Score: Music as Performance*, Oxford: Oxford University Press, 2013.

Curzon, George Nathaniel, *Modern Parliamentary Eloquence, London: Macmillan and co*, 1914.

Damousi, Joy, *Colonial Voices: A Cultural History of English in Australia 1840–1940*, Cambridge, UK: Cambridge University Press, 2010.

Davidson, Hilary, "The Embodied Turn: Making and Remaking Dress as an Academic Practice," *Fashion Theory*, 23:3 (2019), pp. 329–362.

Davies, James Q., *Romantic Anatomies of Performance*, Berkeley: University of California Press, 2014.

Davies, James Q. "Pneumotypes: Jean de Reszke's High Pianissimos and the Occult Sciences of Breathing," in: Benjamin Walton and David Trippett (eds.), *Nineteenth Century Opera and the Scientific Imagination*, Cambridge, UK: Cambridge University Press, 2019.

Davies, James Q., "The Sontag-Malibran Stereotype," *Romantic Anatomies of Performance*, Oakland: University of California Press, 2014, pp. 66–92.

De Bont, Raf, "'Writing in Letters of Blood': Manners in Scientific Dispute in Nineteenth-Century Britain and the German Lands," *History of Science*, 51:3 (2013), pp. 309–335.

Delaere, Mark, "De toekomst van de muziekwetenschap: wat is een partituur?," *Tijdschrift voor Muziektheorie*, 11:1 (2006), pp. 31–35.

Delph Janiurek, Tom, "Sounding Gender(ed): Vocal Performances in English University Teaching Spaces," *Gender, Place and Culture: A Journal of Feminist Geography*, 6: 2, (1999), pp. 137–153.

Derrida, Jacques, "The Voice That Keeps Silence," in: Jonathan Sterne (ed.), *The Sound Studies Reader*, London: Routledge, 2012, pp. 495–504.

Dixon, Thomas, *The Invention of Altruism: Making Moral Meanings in Victorian Britain*, Oxford: Oxford University Press, 2008.

Dolar, Mladen, *A Voice and Nothing More*, Cambridge, MA: MIT Press, 2006.

Du Bois, W. E. B. *The Souls of Black Folk*, New York: Penguin, 1903.

Duden, Barbara, *Geschichte unter der Haut: ein Eisenacher Arzt und seine Patientinnen um 1730*, Eberbach: Greif-Bücher, 1987.

Eidsheim, Nina Sun, *The Race of Sound: Listening, Timbre and Vocality in African American Music*, Durham, NC: Duke University Press, 2018.

Elsberger, Manfred, *Nina d'Aubigny von Engelbrunner: Eine adelige Musikpädagogin am Übergang vom 18. zum 19. Jahrhundert*, BUCH and medi@, 2000.

Enns, Anthony, "The Human Telephone: Physiology, Neurology and Sound Technologies," in: Daniel Morat (ed.), *Sounds of Modern History: Auditory Cultures in 19th- and 20th-Century Europe*, New York: Berghahn, 2014.

Farge, Arlette, *Essai pour une histoire des voix aux dix-huitième siècle*, Paris: Bayard, 2009.

Feldman, Martha, and Judith T. Zeitlin, *The Voice as Something More: Essays Toward Materiality*, Chicago: University of Chicago Press, 2019.

Feldman, Martha, Emily Wilbourne, Steven Rings, Brian Kane, and James Q. Davies, "Why Voice Now?," *Journal of the American Musicological Society*, 68:3 (2015), pp. 653–685.

Feldman, Martha, *The Castrato. Reflections on Natures and Kinds*, Oakland: University of California Press, 2015.

Frevert, Ute, *Gefühle in der Geschichte. Kritische Studien zur Geschichtswissenschaft*, Göttingen: Vandenhoeck & Ruprecht, 2021.

Gardey, Delphine, *Écrire, Calculer, Classer. Comment une révolution de papier a transformé les sociétés contemporaines (1800–1940)*, Paris: Éditions la découverte, 2008.

Gardey, Delphine, *Le linge du Palais Bourbon. Corps, materialité et genre du politique á l'ère démocratique*, Lormont: Éditions Le Bord de l'Eau, 2015.

Gieryn, Thomas F., *Cultural Boundaries of Science: Credibility on the Line*, Chicago: University of Chicago Press, 1999.

Gitelman, Lisa, *Paper Knowledge: Toward a Media History of Document*, Durham, NC: Duke University Press, 2014.

Goldhill, Simon, "The Value of Greek: Why Save Plutarch," in: *Who Needs Greek? Contests in the Cultural History of Hellenism*, Cambridge, UK: Cambridge University Press, 2002.

Goltermann, Svenja, *Körper der Nation. Habitusformierung und die Politik des Turnens, 1860–1890*, Göttingen: Vandenhoeck & Ruprecht, 1998.

Green, Lucy, *Music, Gender and Education*, Cambridge, UK: Cambridge University Press, 1997.

Griffin, Ben, *The Politics of Gender in Victorian Britain: Masculinity, Political Culture, and the Struggle for Women's Rights*, Cambridge, UK: Cambridge University Press, 2012.

Grotjahn, Rebecca, "Seelenvolle Maschine: Natur und Technik im Gesangsdiskurs," in: Vera Grund, Claire Genewein and Hans Georg Nicklaus (eds.), *Naturalezza | Simplicité: Natürlichkeit im Musiktheater*, Bielefeld: Transcript, 2019, pp. 81–107.

Haapala, Taru, *Political Rhetoric in the Oxford and Cambridge Unions, 1830–1870*, London: Palgrave Macmillan, 2016.

Hakosalo, Heini, *On Speaking Terms: Scientific Boundary Work and the Discovery of Aphasia, 1861–1874*, Oulu: Doctoral dissertation, 2006.

Haraway, Donna J. *Modest_Witness@Second_Millennium. FemaleMan©_Meets_OncoMouse^a: Feminism and Technoscience*, New York: Routledge, 1997.

Haraway, Donna, "A Cyborg Manifesto: Science, Technology, and Socialist-Feminism in the Late Twentieth Century," in: *Simians, Cyborgs and Women: The Reinvention of Nature*, New York: Routledge, 1991, pp. 149–181.

Harris, Anna, and Tom Rice, *Stethoscope: The Making of a Medical Icon*, London: Reaktion Books, 2022.

Hartley, Lucy, "Constructing the Common Type: Physiognomic Norms and the Notion of 'Civic Usefulness', from Lavater to Galton," in: Waltraud Ernst (ed.), *Histories of the Normal and the Abnormal: Social and Cultural Histories of Norms and Normativity*, New York: Routledge, 2006, pp. 101–121.

Hawkins, Stan (ed.), *Queerness in Pop Music. Aesthetics, Gender Norms, and Temporality*, London: Routledge, 2016.

Heath, John, *The Talking Greeks: Speech, Animals, and the Other in Homer, Aeschylus, and Plato*, Cambridge, UK: Cambridge University Press, 2005.

Heckendorn Cook, Elizabeth, *Epistolary Bodies: Gender and Genre in the 18th Century Republic of Letters*, Redwood City, CA: Stanford University Press, 1996.

Hewitt, Martin, *Making Social Knowledge in the Victorian City: The Visiting Mode in Manchester, 1832–1914*, London: Routledge, 2020.

Hey, Julius, *Deutscher Gesangs-Unterricht,* Mainz: B. Schotts Söhne, 1884.

Heyer, Anne, and Anne Petterson, "Learning from the Outside: Parliament's Response to Public Meetings in Germany and the Netherlands, 1870–1914," *Parliaments, Estates and Representation*, 44:1, (2024), 64–79.

Himmelfarb, Gertrude, "Manners into Morals: What the Victorians Knew," *American Scholar,* 57:2 (1988), 223–232.

Hoegaerts, Josephine, "«Acquiring a 'Manly' Tone»: Audible Intersections of Masculinity and Age in the Long Nineteenth Century," *Genesis: revista della società italiana delle storiche,* 19:2 (2020), pp. 65–87.

Hoegaerts, Josephine, "Little Citizens and Petites Patries: Learning Patriotism Through Choral Singing in Antwerp in the Late Nineteenth Century," in: Ursula Geisler and Karin Johansson (eds.), *Choral Singing: Histories and Practices*, Cambridge, UK: Cambridge Scholars Publishing, 2014, pp. 14–32.

Hoegaerts, Josephine, "Recording the Subaltern's Speech. Children's Voices in the Antwerp School Archives, ca. 1850–1900," *Journal of Belgian History*, 46:1 (2016), pp. 10–31.

Hoegaerts, Josephine, "Speaking Like Intelligent Men: Vocal Articulations of Authority and Identity in the House of Commons in the Nineteenth Century," *Radical History Review* (themed issue: Sound Politics), 121 (2015), pp. 123–144.

Hoegaerts, Josephine, "The Sound of Sovereignty: Royal Vocal Strategies in the Victorian House of Lords," in: Anke Gilleir and Aude Defurne (eds.), *Strategic Imaginations: Women and the Gender of Sovereignty in European Culture*, Leuven: Leuven University Press, 2020.

Hoegaerts, Josephine, "By but Not of the Body: Performing Girls' and Boys' Voices in the Nineteenth Century," in: Heidrun Zettelbauer and Stefan Benedik (eds.), *Embodiment–Verkörperung*, Göttingen: Vandenhoeck & Ruprecht, 2017.

Hoegaerts, Josephine, "Women's Voices in Educational Manuals. The Gendered Sounds of Speech Therapy, Song and Education in Europe, c. 1830–1900," *Women's History Review* (2019).

Hoegaerts, Josephine, and Janice Schroeder (eds.), *Ordinary Oralities: Everyday Voices in History*, Oldenbourg: De Gruyter, 2023.

Höfer, Bernadette, *Psychosomatic Disorders in Seventeenth-Century French Literature*, Aldershot: Ashgate, 2009.

Huisman, Frank, "Shaping the Medical Market. On the Construction of Quackery and Folk Medicine in Dutch Historiography," *Medical History*, 43:3 (1999), pp. 359–375.

Hunter, John, *The Spirit of Self-Help: a Life of Samuel Smiles*, London: Shepheard-Walwyn, 2017.

Hurd, Madeleine, "Class, Masculinity, Manners and Mores. Public Space and Public Sphere in Nineteenth-Century Europe," *Social Science History*, 24:1 (2000), pp. 75–101.

Jackson, Sarah, "Derrida on the Line," *Derrida Today*, 10:2 (2017), pp. 142–159.

Jacyna, L. S., "The 1874 Aphasia Debate in the *Berliner Gesellschaft für Anthropologie*," *Brain and Language*, 69 (1999), pp. 5–15.

James, Frank A. J. L., "'Never Talk About Science, Show It to Them': The Lecture Theatre of the Royal Institution," *Interdisciplinary Science Reviews*, 27:3 (2002), pp. 225–229.

Jarman-Ivens, Freya, *Queer Voices: Technologies, Vocalities and the Musical Flaw*, London: Palgrave-Macmillan, 2011.

Jung, Theo, "Le Silence du Peuple: The Rhetoric of Silence During the French Revolution," *French History*, 31:4 (2017), pp. 440–469.

Kafka, Franz, "A Report for an Academy," translated by Ian Johnston. Original: Franz Kafka, "Ein Bericht für eine Akademie," *Der Jude. Eine Monatsschrift*. Jg. 2, Heft 8., 1917.

Kane, Brian, *Sound Unseen: Acousmatic Sound in Theory and Practice*, Oxford: Oxford University Press, 2014.

Kivimäki, Ville, Tanja Vahtikari, and Sami Suodenjoki, *Lived Nation: Histories of Experience and Emotion in understanding Nationalism*, London: Palgrave, 2020.

Knight, David, "Scientific Lectures: A History of Performance," *Interdisciplinary Science Reviews*, 27:3 (2002), pp. 217–224.

Koestenbaum, Wayne, *The Queen's Throat: Opera, Homosexuality and the Mystery of Desire*, New York: Da Capo Press, 1993.

Kocka, Jürgen, "Bürgertum und Bürgerlichkeit als Probleme der deutschen Geschichte vom späten 18. zum frühen 20. Jahrhundert," in: J. Kocka (ed.), *Bürger und Bürgerlichkeit im 19. Jahrhundert*, Göttingen: Vandenhoeck & Ruprecht, 1987.

Kohler, Robert E. "Lab History: Reflections," *Isis*, 99:4 (2008), pp. 761–768.

Kreilkamp, Ivan, *Voice and the Victorian Storyteller*, Cambridge, UK: Cambridge University Press, 2005.

Kuchta, David, "The Making of the Self-Made Man," in: *The Three-Piece Suit and Modern Masculinity: England, 1550–1850*, Oakland: University of California Press, 2002.

Lacey, Kate, "Speaking Up and Listening Out: Media Technologies and the Re-Sounding of the Public Sphere," in: Dmitri Zakharine and Nils Meise (eds.), *Electrified Voices: Medial, Socio-Historical and Cultural Aspects of Voice Transfer*, Göttingen: Vandenhoeck und Ruprecht, 2013, pp. 123–136.

Lane, Harlan, *The Mask of Benevolence: Disabling the Deaf Community*, San Diego: Dawn Sign Press, 1992.

Laqueur, Thomas, *Making Sex: Body and Gender from the Greeks to Freud*, Cambridge, MA: Harvard University Press, 1992.

Lawlor, Leonard, "Auto-Affection," in: *Derrida: Key Concepts*, Routledge, 2014, pp. 130–138.

Le Breton, André, *Du Silence*, Paris: Metallié, 1997.

Lebrun, Yvan, and Madeleine Bayle, "Surgery in the Treatment of Stuttering," in: Yvan Lebrun and Richard Hoops (eds.), *Neurolinguistic Approaches to Stuttering*, Berlin: Mouton, 1973, pp. 82–90.

Lorch, Marjorie, "The Long View of Language Localization," *Frontiers in Neuroanatomy*, 13:52 (2019).

Luckyj, Christina, *"A Moving Rhetoricke": Gender and Silence in Early Modern England,* Manchester: Manchester University Press, 2002.

Lurton, Guillaume, *Le Chœur partagé. Le chant choral en France, intégration socio-économique d'un monde de l'art moyen*, doctoral thesis, Sociologie. Sciences Po—Institut d'études politiques de Paris, 2011.

Marantz Cohen, Paula, *Talking Cure: An Essay on the Civilizing Power of Conversation*, Princeton: Princeton University Press, 2023.

Martensen, Karin, *"The Phonograph Is Not an Opera House": Quellen und Analysen zu Ästhetik und Geschichte der frühen Tonaufnahme am Beispiel von Edison und Victor*, Munich: Allitera Verlag, 2019.

Martin, Daniel, "George Catlin's Shut Your Mouth, the Biopolitics of Voice, and the Problem of the 'Stuttering Indian,'" in: Josephine Hoegaerts and Janice Schroeder (eds.), *Ordinary Oralities. Everyday Voices in History*, Oldenbourg: De Gruyter, 2023, pp. 65–80.

Martin, Daniel, "Speaking Machines and Ghostly Phantoms: The Claustrum Poetics of Voice and Dysfluency," *Victorian Review*, 46:1 (2020), pp. 1–5.

Mattfeld, Monica, *Becoming Centaur: Eighteenth-Century Masculinity and English Horsemanship*, University Park: Pennsylvania State University Press, 2016.

McGuire, Riley, "The Victorian Unspeakable: Stammering and Same-Sex Intimacy Between Men." *DiGeSt. Journal of Diversity and Gender Studies* 3:2 (2016), pp. 43–57.

Mecke, Ann Christine, *Mutantenstadl. Der Stimmwechsel und die deutsche Chorpraxis im 18. und 19. Jahrhundert*, Berlin: Wissenschaftlicher Verlag Berlin, 2007.

Meisel, Joseph, *Public Speech and the Culture of Public Life in the Age of Gladstone*, New York: Columbia University Press, 2001.

Milam, Erika Lorraine and Robert Nye (eds.), "Scientific Masculinities (special issue), *Osiris*, 30: 1 (2015).

Miller-Frank, Felicia, *The Mechanical Song: Women, Voice, and the Artificial in Nineteenth-Century French Narrative*, Redwood, CA: Stanford University Press, 1995.

Morrison, Toni, *Song of Solomon*, New York: Alfred Knopf, 1977.

Morus, Iwan Rhys, *Shocking Bodies: Life, Death & Electricity in Victorian England*, Cheltenham: History Press, 2011.

Mullender, Rob, "Divine Agency: Bringing to Light the Voice Figures of Margaret Watts-Hughes," *Sound Effects*, 8:1 (2019), pp. 123–139.

Nelson, Ingrid L., "Interspecies Care and Aging in a Gorilla 2.0 World," *Geoforum*, 79 (2017), pp. 144–152.

Nicklaus, Hans Georg, "Stimmen der Unschuld. Kastratengesang im Kontext von Kindheit und Pädagogik," in: Vera Grund, Claire Genewein and Hans Georg Nicklaus (eds.), *Naturalezza | Simplicité: Natürlichkeit im Musiktheater*, Bielefeld: Transcript, 2019.

Noa, Miriam, *Volkstümlichkeit und Nationbuilding. Zum Einfluss der Musik auf den Einigungsprozess der deutschen Nation im 19. Jhdt*, New York: Waxman, 2013.

Ochoa Gautier, Ana Maria, *Aurality: Listening and Knowledge in Nineteenth-Century Colombia*, Durham, NC: Duke University Press, 2014.

Ohala, John J., "Christian Gottlieb Kratzenstein: Pioneer in Speech Synthesis," *ICPhS*, 17 (2011), pp. 156–159.

Ong, Walter, "Writing Is a Technology that Restructures Thought," in: Pamela Downing, Susan Lima, Michael Noonan, J. Benjamins (eds.), *The Linguistics of Literacy*, 1992, pp. 293–319.

Ong, Walter, *Orality and Literacy: The Technologizing of the Word*, London: Methuen & Co., 1982.

Ottosson, Anders, "The First Historical Movements of Kinesiology: Scientification in the Borderline Between Physical Culture and Medicine Around 1850," *International Journal of the History of Sport*, 27:11 (2010), pp. 1892–1919.

Paige, Nicholas D., *Being Interior: Autobiography and the Contradiction of Modernity in Seventeenth-Century France*, Philadelphia: University of Pennsylvania Press, 2000.

Pemberton, Cecilia, Paul McCormack, and Allison Russell, "Have Women's Voices Lowered Across Time? A Cross Sectional Study of Australian Women's Voices," *Journal of Voice*, 12:2 (1998), 208–213.

Picker, John, "The Recorded Voice from Victorian Aura to Modernist Echo," *Victorian Soundscapes*, Oxford: Oxford University Press, 2003, pp. 110–146.

Pinch, Trevor, and Karin Bijsterveld (eds.), *The Oxford Handbook of Sound Studies*, Oxford: Oxford University Press, 2012.

Pinch, Trevor, and Karin Bijsterveld, Introduction to "Sound Studies: New Technologies and Music," Special Issue, *Social Studies of Science*, 34:5 (2004).

Pitman, Isaac, *Shorthand Phonography*, London: Isaac Pitman and Sons, 1837

Poizat, Michel, *L'Opera, ou, Le cri de l'ange: Essai sur la jouissance de l'amateur d'opera*, Paris: Métailié, 1986.

Potter, John, "The Tenor–Castrato Connection, 1760–1860," *Early Music*, 35 (2007), pp. 97–112.

Potter, John, and Neil Sorrell, *A History of Singing*, Cambridge, UK: Cambridge University Press, 2012.

Potter, John, *Tenor: History of a Voice*, New Haven, CT: Yale University Press, 2009.

Potter, John, *Vocal Authority: Singing Style and Ideology*, Cambridge, UK: Cambridge University Press, 2006.

Puwar, Nirmal, *Space Invaders: Race, Gender and Bodies out of Place*, Oxford: Berg Publishers, 2004.

Quintilian, *Institutione Oratoria*, edited by Harold Edgeworth Butler, *Perseus Digital Library 4.0*, https://www.perseus.tufts.edu/hopper/.

Reddick, Allan, "Recording the Most Proper and Significant Words," in: Sarah Ogilvie (ed.), *Eighteenth-Century English Dictionaries: Prescriptivism and Completeness*, Cambridge, UK: Cambridge University Press, 2020.

Rée, Jonathan, *I See a Voice. Deafness, Language and the Senses*, London: Metropolitan Books, 1999.

Reid, Christopher, *Imprison'd Wranglers: The Rhetorical Culture of the House of Commons 1760–1800*, Oxford: Oxford University Press, 2012.

Ribac, François, and Viktoria Tkaczyk. "Research on the History of Modern Acoustics," *Revue d'Anthropologie des Connaissances*, 13:3 (2019), pp. 707–720.

Richards, Jennifer, *Voices and Books in the English Renaissance: A New History of Reading*, Oxford: Oxford University Press, 2019.

Richardson, Sarah, *The Political Worlds of Women: Gender and Politics in 19th-Century Britain*, New York: Routledge, 2013.

Richter, *Demokratie. Eine deutsche Affäre*, Munich: C.H. Beck, 2020.

Robertson-Kirkland, Brianna E., "The Silencing of Bel Canto," *eSharp*, 21 (2013).

Robinson, Dylan, *Hungry Listening: Resonant Theory for Indigenous Sound Studies*, Minneapolis: University of Minnesota Press, 2020.

Rockey, Denyse, *Speech Disorder in Nineteenth Century Britain: The History of Stuttering*, London: Groom Helm, 1980.

Rossing, Thomas, "A Brief History of Acoustics," in Thomas Rossing (ed.), *Handbook of Acoustics*, New York: Springer, 2007, pp. 9–24.

Ruberg, Willemijn, *History of the Body. Theory and History*, London: Bloomsbury Academic, 2020.

Russell, Dave, *Popular Music in England 1840–1914: A Social History*, Manchester: Manchester University Press, 1997.

Rutherford, Susan, *The Prima Donna and Opera, 1815–1930*, Cambridge, UK: Cambridge University Press, 2006.

Siebs, Theodor, *Deutsche Bühnenausspache*, Bonn: Verlag Albert Ahn, 1898.

Sarasin, Philip, *Anthrax. Bioterror als Phantasma*, Frankfurt a.M: Suhrkamp, 2004.

Schweitzer, Claudia, "La voix humaine: instrument à vent ou instrument à cordes? Un parcours historique (XVIIe/XVIIIe siècles) autour de cette question," *Language & History*, 63:3 (2020).

Sconce, Jeffrey, *Haunted Media: Electronic Presence from Telegraphy to Television*, Durham, NC: Duke University Press, 2000.

Shapin, Steven, and Simon Schaffer, *Leviathan and the Air-Pump: Hobbes, Boyle, and the Experimental Life*, Princeton, NJ: Princeton University Press, 1985.

Shigo, Daniel, *Voice Talk*, "The Garcia Method and Singing in the Mask," https://www.shigovoicelessons.com/voicetalk//2014/07/the-garcia-method-singing-in-mask.html

Smith, Bruce R., *The Acoustic World of Early-Modern England*, Chicago: University of Chicago Press, 1999.

Solanki, Tanvi, "Listening to Difference: J.G. Herder's Aural Theory of Cultural Diversity in the 'Treatise on the Origin of Language' (1772)," *History of European Ideas*, 48:7 (2022), pp. 930–947.

Sparrow, Andrew, *Obscure Scribblers: A History of Parliamentary Journalism*, London: Politico, 2003.

St. Pierre, Joshua, and Charis St. Pierre, "Governing the Voice: A Critical History of Speech-Language Pathology," *Foucault Studies*, 24 (2018), pp. 151–184.

St. Pierre, Joshua, *Cheap Talk: Disability and the Politics of Communication*, Ann Arbor: University of Michigan Press, 2022.

Stark, James, *Bel Canto: A History of Vocal Pedagogy*, Toronto: University of Toronto Press, 1999.

Sterne, Jonathan, *MP3, the Meaning of a Format*, Durham, NC: Duke University Press, 2012.

Sterne, Jonathan, *The Audible Past: Cultural Origins of Sound Reproduction*, Durham, NC: Duke University Press, 2003.

Stevens, Elizabeth, *Anatomy as Spectacle: Public Exhibitions of the Body from 1700 to the Present*, Liverpool: Liverpool University Press, 2011.

Stigler, Stephen, *The History of Statistics: The Measurement of Uncertainty Before 1900*, Cambridge, MA: Belknap Press, 1986.

Stoever, Jennifer Lynn, *The Sonic Color Line: Race and the Cultural Politics of Listening*, New York: NYU Press, 2016.

Supper, Alexandra, and Karin Bijsterveld, "Sounds Convincing: Modes of Listening and Sonic Skills in Knowledge Making," *Interdisciplinary Science Reviews*, 40 (2015), pp. 124–144.

Sykes, Ingrid, "Sounding the Citizen Patient: The Politics of Voice in Post-Revolutionary France," *Medical History*, 55 (2011), pp. 479–502.

te Velde, Henk, *Sprekende politiek. Redenaars en hun publiek in de gouden eeuw*, Amsterdam: Prometheus, 2015.

The Granta, February 2nd, 1895.

Thurman, Kira, *Singing Like Germans: Black Musicians in the Land of Bach, Beethoven, and Brahms*, Ithaca, NY: Cornell University Press, 2021.

Tkaczyck, Viktoria, "Whose Larynx Is It? Fields of Scholarly Competence around 1900," *History of Humanities*, 3:1 (2018), pp. 57–73.

Trower, Shelley, *Senses of Vibration: A History of the Pleasure and Pain of Sound*, London: Continuum, 2012.

Vandendriesche, Joris, Evert Peeters, and Kaat Wils, "Performing Expertise," in: Joris Vandendriessche, Evert Peeters, and Kaat Wils (eds.), *Scientists' Expertise as Performance*, London: Routledge, 2015.

Vandendriessche, Joris, "Setting Scientific Standards: Publishing in Medical Societies in Nineteenth-Century Belgium," *Bulletin of the History of Medicine*, 88:4 (2014), pp. 626–653.

von Kempelen, Wolfgang, *Mechanismus der menschlichen Sprache: nebst der Beschreibung seiner sprechenden Maschine*, J. V. Degen, 1791.

von Polenz, Peter, *Deutsche Sprachgeschichte vom Spätmittelalter bis zur Gegenwart*, Berlin: Degruyter, 2000.

Voskuhl, Adelheid, *Androids in the Enlightenment: Mechanics, Artisans and Cultures of the Self*, Chicago: University of Chicago Press, 2013.

Walker, Oriana, *The Breathing Self: Toward a History of Respiration*. Doctoral dissertation, Harvard University, 2016.

Wierzbicka, Anna, *Experience, Evidence and Sense: The Hidden Cultural Legacy of English*, Oxford: Oxford University Press, 2010.

Wils, Kaat, Mineke Bosch, and Kirstin Niskanen (eds.), Special Issue: Scientific Persona, *Persona Studies*, 4:1 (2018).

Wilson Kimber, Marian, *The Elocutionists: Women, Music, and the Spoken Word*, Champaign: University of Illinois Press, 2017.

Wright, Tom, *Lecturing the Atlantic: Speech, Print, and an Anglo-American Commons 1830–1870*, Oxford: Oxford University Press, 2017.

Yamaguchi, Liesl, "Sensuous Linguistics: On Saussure's Synaesthesia," *New Literary History*, 50:1 (2019), pp. 23–42.

Yasar, Kerim, *Electrified Voices: How the Telephone, Phonograph, and Radio Shaped Modern Japan, 1868–1945*, New York: Columbia University Press, 2018.

Zakharine, Dmitri, "Preface: Electrified Voices. Medial, Socio-Historical and Cultural Aspects of Voice Transfer in Dmitri Zakharine and Nils Meise (eds.), *Electrified Voices: Medial, Socio-Historical and Cultural Aspects of Voice Transfer*, Göttingen: V&Runpress, 2013, 12–29.

INDEX

ACKNOWLEDGMENTS

This book has been a long time in the making and I have amassed a multitude of debts of gratitude (and have owed people an impressive number of drinks) over the years. The first seeds for the idea of this study in the history of the sounding voice were sown at the University of Leuven, where, funded by an FWO grant, I had a chance to start my archival work on the scientific history of laryngology. In subsequent years, I worked on related projects of vocal education and political speech as a visiting scholar at Birkbeck College, the Hochschule für Musik und Theater Félix Mendelssohn Bartholdy, the University of the Aegean, and the Cambridge Centre for Research in the Arts, Humanities and Social Sciences, funded by FWO, the Helsinki University Future Fund, and the European Research Council, respectively. These travels gave me a chance not only to visit some beautifully managed archives, but also to rely on the considerable expertise of their staff. I'm immensely grateful to the many wonderful librarians and archivists at the Bibliothèque National de Paris, the Archives Nationales at Pierrefitte, the Samuel Heinicke Bibliothek, the libraries of the Hochschule für Musik und Theater Félix Mendelssohn Bartholdy, of the Royal College of Music and the Royal Academy of Music, and the Wellcome Collection, who have supported me in my search for obscure scientific essays, advertisements for miracle cures, and eccentric singing methods.

Most of my time, however, I have spent among a range of marvelous colleagues teaching and researching alongside me at the Universities of Helsinki and Amsterdam. The crew of CALLIOPE, Ludovic Marionneau, Karen Lauwers, Esha Sil, and Liesl Yamaguchi, have been a wonderful sounding board, whose shared expertise on political and poetic speech has shaped my thinking around the societal salience of vocal practice. The reading group at the Helsinki Institute for Advanced Studies, consisting of Eva Johanna Holmberg, Katja Ritari, Samuli Hurre, Nora Hämäläinen, Merja Polvinen, Jan Stenger, Thomas Devaney, and Ulrika Maud, provided a remarkably wide range of feedback on the early iterations of the arguments presented here. The scholars gathered in the Amsterdam

School for Regional, Transregional and European Studies as well as the members of the chair group for European Culture, Yolanda Rodriguez-Perez, Alex Drace-Francis, Guido Snel, Marleen Rensen, Krisztina Lajosi, Marjet Brolsma, Linde Luijnenburg, Jesse van Amelsvoort, Nuri Kurnaz, Carlijn Cobers, Francesca Zantedeschi, Sudha Rajagopalan, Matthijs Lok, Robin De Bruin, Ewa Stanczyk, Theresa Kuhn, Micha Kemper, and many others, have helped me polish my arguments as I was concluding my work on the book. Editors Emma Dillon and Walter Biggins had a large hand in turning my—by then sprawling—research into a readable book, offering many useful suggestions toward its coherence. Most of all, they gave the book a home in the lovely *Sound in History* series. Simon Turner, finally, has applied their finely tuned ear to the manuscript in its final stages, and I am grateful for the elegant sensitivity they have brought to editing my eccentric turn of phrase.

In between, I was lucky to be able to discuss various chapters with numerous colleagues and graduate students in workshops and seminars at the University of Oulu, the University of Stockholm, the Sibelius Academy, the University of Bologna, the University of the Aegean, the University of Tampere, and the Cambridge Cultural History Seminar. Special thanks goes to the Helsinki Seminar for Cultural Studies and Cultural History, and especially my co-convenors Anu Korhonen and Outi Hakola, as well as my graduate students Sofie Gregersen, Jana Lainto, Clarice Sääväla, Vincent Veerbeek, and Lotta Vuorio, for not only reading and commenting on various iterations of the manuscript with enthusiasm, but also drawing me into the orbit of their own amazing work. They have broadened my horizons in no small measure.

I have been the recipient of a great deal of generosity by scholars who not only have helped me sharpen my thinking by reading chapters and letting me pick their impressive brains, but have also become valued friends: Daniel Blackie, Jan Bleyen, Ben Griffin, Sasha Grigorieva, Freya Jarman, Sophie Holm, Jan Schroeder, Riikka Taavetti, Henning Trüper, Joris Vandendriessche, and Pieter Verstraete have all contributed to the ideas in this book, probably in more ways than they realize. I hope that my little club of strong-minded Helsinki friends and colleagues, Elizabeth Peterson, Hanna Snellman, Suzie Thomas, and Xenia Zeiler, already know how invaluable they have been to me over the years, and that they may find their fierce support reflected here as well.

Then there are those to whom I owe more than I will probably ever be able to repay . . . the friends and family who have patiently let me blather on about eidophones, commiserated with every disappointment, and rejoiced over every success in the more precarious periods in which this book was also written, who endured long phone calls when pandemics and distance allowed for nothing else, and who dragged me out of the library and into the world. Marc, Anne-Marie, Wouter, Dolf, Veerle, Edith, Roel, Nikkie, Michael, Allan, Anna, and Nico: in so many ways, this book would never have seen the light of day without you.

Finally, so much of what has driven my interest in the sensorial experience and emotional expressiveness of the human voice has been the experience of being surrounded by so many of them, and especially by some of its most beautiful specimens. Having sung—well, everywhere, but especially—in choirs since I was a little girl, I know that it is often the warmth of a human voice that will get me through the day. During the writing of this book, I was happily ensconced within the harmonies and occasional dissonances of Vocaal Ensemble Markant, Capella Sanctorum Michaelis et Gudulae, Ensemble Vocal Amaryllis, the Choer Symphonique de Liège, the Camden Choir, Helsinki Yliopiston Musiikkiseuran Kamarikuori, Akateeminen Laulu, Ceciliakören, Ralph Sordo, Solderkoor, and Lelikoor. Apart from teaching me more about singing than all the handbooks I consulted for this research combined, they have been places to find and tend to some of my most nourishing and enduring friendships.

My fellow alto, former choir neighbor, co-instigator of many an ill-advised concert, and, above all, my staunch friend, Griet, is one of the steadiest voices in my life, as well as mother to one of my favorite duets, which includes my spirited godson, Gaspard. This book is dedicated to them.

www.ingramcontent.com/pod-product-compliance
Lightning Source LLC
LaVergne TN
LVHW091527200926
848483LV00002B/326